Working With Numbers

Refresher

AUTHOR
James T. Shea

CONSULTANT
Susan L. Beutel
Consulting/Resource Teacher
Lamoille North Supervisory Union, VT

ACKNOWLEDGMENTS

Executive Editor: Wendy Whitnah

Senior Math Editor: Donna Rodgers

Design Coordinator: John Harrison

Project Design
and Development: The Wheetley Company

Cover Design and
Electronic Production: John Harrison, Adolph Gonzalez and
Chuck Joseph

WORKING WITH NUMBERS SERIES:

Level A	Level D	Consumer Math
Level B	Level E	Refresher
Level C	Level F	Algebra

ISBN: 0-8114-5224-7

STECK-VAUGHN
COMPANY
ELEMENTARY • SECONDARY • ADULT • LIBRARY

TABLE OF CONTENTS

3

TO THE LEARNER

This book provides you with lessons that teach basic mathematical skills. A step-by-step procedure for solving problems is followed by practice exercises. The lessons provide the drill and practice needed for you to master and remember the skills. Also included in the book are lessons on problem-solving strategies that will help you to apply the skills to your everyday life.

This book contains a number of features that will help you master the mathematics that you will need in your everyday life. Among these are the following:

A Pretest and a Mastery Test

At the beginning of the book, the Pretest will tell you what material you may know already and what material you have not yet mastered.
At the end of the book, the Mastery Test will tell you what material you have learned and what you may need to review further.

Unit Reviews

Each unit concludes with a Unit Review that provides you with an opportunity to demonstrate your understanding of the skills and concepts presented in that particular unit.

Problem-Solving Strategies

A problem-solving strategy is an effective plan for solving a problem. There are two lessons in each unit on problem-solving strategies. So by the time you finish the book, you will have learned sixteen different strategies for solving problems.
There also are at least two problem-solving lessons in each unit where you have the opportunity to choose strategies to solve problems.

Answers and Some Solutions

The answers to all of the problems are provided in the back of the book. This allows you to check your work after it is completed. Solutions, or step-by-step explanations, are provided for selected problems. These take you through the steps used to solve the problems.

PRETEST

Compare. Write >, <, or =.

	a	b	c
1.	43 _____ 28	416 _____ 495	3060 _____ 3600
2.	0.36 _____ 0.036	0.9 _____ 0.90	2.39 _____ 2.4

Write in order from least to greatest.

3. 135   516   398 _____

4. 48.6   0.486   40.9 _____

Write the value of the underlined digit.

	a	b	c	d
5.	8̲43 _____	2̲6,498 _____	41.9̲4 _____	8.3̲92 _____

Round to the nearest hundred.

6. 362 _____   841 _____   750 _____   499 _____

Add.

	a	b	c	d	e
7.	4	3	4 2 6	7 4 2	5.0 6
	2	8	3 8 5	1 8	2 0.7 1 6
	+ 3	7	2 4 0	3 6 4 7	2 4.9
		+ 5	+ 1 6 2	+ 8 9 2	+ 8.4 7

Name _____   Date _____

Find each answer.

	a	b	c	d
8.	483 + 215	3146 + 4832	4968 + 3217	56,295 + 7,483
9.	485 − 354	8926 − 2315	306 − 298	6037 − 3647
10.	93 × 2	23 × 31	87 × 36	485 × 346
11.	5 $)\overline{78}$	4 $)\overline{562}$	5 $)\overline{429}$	42 $)\overline{985}$

Line up the digits. Then find each answer.

	a	b	c
12.	846 − 97 = _____	495 + 632 = _____	52,631 − 894 = _____
13.	24 × 367 = _____	553 ÷ 6 = _____	4271 ÷ 54 = _____

Simplify.

	a	b	c	d	e
14.	$\frac{24}{36} =$	$\frac{28}{5} =$	$\frac{5}{35} =$	$\frac{44}{9} =$	$\frac{13}{39} =$

Solve.

15. Jan needs a sheet of paper that is $8\frac{3}{8}$ inches long. She has a piece that is 11 inches long. How much must she cut off?

Answer _____

16. Carpet sells for $36 per square yard. How much will $8\frac{3}{4}$ square yards cost?

Answer _____

Add or subtract. Simplify. Watch the operation signs.

	a	b	c	d
17.	$\dfrac{5}{8}$ $+ \dfrac{1}{8}$	$\dfrac{5}{9}$ $+ \dfrac{7}{9}$	$\dfrac{3}{4}$ $+ \dfrac{5}{12}$	$\dfrac{1}{4}$ $+ \dfrac{5}{6}$
18.	$\dfrac{4}{7}$ $- \dfrac{2}{7}$	$\dfrac{2}{3}$ $- \dfrac{2}{6}$	$3\dfrac{5}{8}$ $+ 4\dfrac{1}{4}$	$5\dfrac{2}{3}$ $- 1\dfrac{1}{2}$
19.	$3\dfrac{3}{5}$ $+ 5\dfrac{5}{6}$	$4\dfrac{1}{5}$ $- 2\dfrac{7}{10}$	$5\dfrac{8}{9}$ $- 4$	8 $- 2\dfrac{5}{9}$

Multiply or divide. Use cancellation when possible. Simplify.

	a	b	c
20.	$\dfrac{9}{10} \times \dfrac{7}{9} =$	$\dfrac{5}{9} \times \dfrac{7}{10} =$	$\dfrac{7}{12} \times \dfrac{4}{9} =$
21.	$\dfrac{5}{6} \times 24 =$	$\dfrac{5}{8} \times 12 =$	$\dfrac{8}{9} \times \dfrac{3}{4} =$
22.	$3\dfrac{4}{5} \times 15 =$	$3\dfrac{3}{4} \times 2\dfrac{1}{5} =$	$2\dfrac{1}{2} \times 3\dfrac{7}{10} =$
23.	$\dfrac{5}{8} \div \dfrac{3}{8} =$	$\dfrac{3}{7} \div 3 =$	$\dfrac{12}{13} \div \dfrac{3}{4} =$
24.	$1\dfrac{3}{4} \div 2\dfrac{1}{2} =$	$3\dfrac{1}{3} \div \dfrac{2}{5} =$	$\dfrac{8}{15} \div 4\dfrac{1}{5} =$
25.	$2\dfrac{1}{3} \div \dfrac{7}{8} =$	$2\dfrac{3}{8} \div 2\dfrac{2}{3} =$	$\dfrac{5}{9} \div \dfrac{1}{5} =$

Name _____ Date _____

Write each decimal as a fraction.

	a	b	c	d
26.	$0.9 =$ _____	$4.65 =$ _____	$0.06 =$ _____	$0.125 =$ _____

Write each fraction as a decimal.

	a	b	c	d
27.	$\frac{7}{10} =$ _____	$\frac{51}{100} =$ _____	$\frac{4}{100} =$ _____	$\frac{5}{1000} =$ _____

Find each answer. Write zeros as needed.

	a	b	c	d
28.	$4.2\,8$ $+\,0.2\,6$	$4\,6.9\,2\,7$ $+\,1\,6.0\,2$	$8.4\,6$ $-\,3.1\,9$	$8\,2.6$ $-\,\;\;5.9\,3$
29.	$7.6\,1$ $\times\;\;\;4\,8$	$0.0\,0\,1\,3\,2$ $\times\;\;\;\;\;\;4\,5\,5$	$5.0\,8$ $\times\;\;\;\;\;7$	$6.3\,4$ $\times\,0.2\,9$
30.	$10\,\overline{)\,1\,6.4}$	$0.3\,\overline{)\,1\,2.0\,3}$	$9\,\overline{)\,4\,1.0\,4}$	$3.4\,\overline{)\,1\,7.8\,5}$

Write each percent as a decimal and as a fraction.

	a	b
31.	$8\% =$ _____ _____	$52\% =$ _____ _____

Find each number.

	a	b
32.	80% of 50	What percent of 36 is 9?
33.	40% of what number is 34?	38% of 240

Solve.

34. How many square inches of glass are needed to cover a picture that measures 9 inches by 12 inches?

Answer _____

35. What is the area of a triangle that has a base of 20 centimeters and a height of 15 centimeters?

Answer _____

9

Name _____   Date _____

Change each measurement to the unit given.

	a	*b*	*c*
36.	15 qt = _____ gal _____ qt	3 pt = _____ c	32 oz = _____ lb
37.	40 in. = _____ ft _____ in.	5 T = _____ lb	4 mL = _____ L
38.	8 kg = _____ g	560 mg = _____ g	6000 g = _____ kg

Solve.

	a	*b*	*c*	*d*
39.	$x + 11 = 40$	$x - 7 = 17$	$9x = 45$	$3x + 5 = 32$
40.	$3x + 4x = 84$	$15x - 3x = 72$	$\frac{3}{x} = \frac{27}{45}$	$\frac{16}{6} = \frac{x}{30}$
41.	$\frac{x}{5} = \frac{20}{25}$	$8x + 6 = 9x + 1$	$4x - 7 = 29$	$8 = 5x - 32$
42.	$\frac{6}{5} = \frac{x}{15}$	$2x - 15 = 31$	$42 + x = 81$	$5x + 12 = 8x$

Write the fraction for each ratio.

	a	*b*
43.	The ratio of 10 women to 8 men	The ratio of days in a week to months in a year.
	Ratio _____	Ratio _____

Solve the proportion.

	a	*b*	*c*
44.	$\frac{7}{8} = \frac{x}{40}$	$\frac{x}{9} = \frac{10}{3}$	$\frac{x}{3} = \frac{12}{18}$

WHERE TO GO FOR HELP

The table below lists the problems in the Pretest and the pages of the book on which the corresponding skills and concepts are taught and practiced. For each problem that you could not answer or answered incorrectly, you can use the table to find the page number or numbers where that skill or concept is taught.

PROBLEMS	PAGES	PROBLEMS	PAGES	PROBLEMS	PAGES	PROBLEMS	PAGES
1a	17	11c	46–47	22c	90	33b	146–147
1b	17	11d	50–51	23a	94–95	34	169
1c	17	12a	29	23b	96–97	35	168
2a	109	12b	21	23c	94–95	36a	170
2b	109	12c	31	24a	101	36b	170
2c	109	13a	42–43	24b	100	36c	102
3	17	13b	46–47	24c	100	37a	102
4	109	13c	50–51	25a	101	37b	102
5a	16	14a	65	25b	101	37c	162
5b	16	14b	63	25c	94–95	38a	160
5c	108	14c	65	26a	110	38b	160
5d	108	14d	63	26b	110	38c	160
6a	18	14e	65	26c	110	39a	178–179
6b	18	15	82	26d	100	39b	180–181
6c	18	16	91	27a	110	39c	178–179
6d	18	17a	66	27b	110	39d	180–181
7a	13	17b	66	27c	110	40a	182–183
7b	13	17c	67	27d	110	40b	182–183
7c	33	17d	68	28a	112–113	40c	191
7d	33	18a	66	28b	112–113	40d	191
7e	113	18b	74	28c	114–115	41a	191
8a	19	18c	69	28d	114–115	41b	184–185
8b	19	18d	78–79	29a	122–123	41c	180–181
8c	21	19a	72	29b	122–123	41d	180–181
8d	21	19b	78–79	29c	122–123	42a	191
9a	28	19c	78–79	29d	124–125	42b	180–181
9b	28	19d	76	30a	130	42c	178–179
9c	29	20a	84–85	30b	127	42d	184–185
9d	29	20b	84–85	30c	126	43a	188
10a	40	20c	84–85	30d	127	43b	188
10b	40	21a	86–87	31a	140	44a	191
10c	41	21b	86–87	31b	141	44b	191
10d	42	21c	84–85	32a	146–147	44c	191
11a	46–47	22a	88	32b	148		
11b	49	22b	90	33a	149		

WHOLE NUMBERS

The Addition Facts

Shown below are some addition facts that you need to know very well.
Practice adding until you have the facts memorized.

PRACTICE

Add.

	a	*b*	*c*	*d*	*e*	*f*	*g*	*h*	*i*	*j*
1.	1 +1	2 +1	0 +1	4 +1	5 +1	7 +1	6 +1	3 +1	9 +1	8 +1
2.	5 +2	1 +2	2 +2	9 +2	7 +2	8 +2	0 +2	6 +2	4 +2	3 +2
3.	3 +3	0 +3	4 +3	1 +3	5 +3	2 +3	9 +3	6 +3	8 +3	7 +3
4.	5 +4	9 +4	7 +4	1 +4	2 +4	6 +4	4 +4	0 +4	3 +4	8 +4
5.	3 +5	5 +5	1 +5	9 +5	4 +5	2 +5	7 +5	6 +5	0 +5	8 +5
6.	6 +6	4 +6	0 +6	5 +6	9 +6	2 +6	8 +6	1 +6	7 +6	3 +6
7.	3 +7	1 +7	5 +7	8 +7	4 +7	7 +7	0 +7	9 +7	2 +7	6 +7
8.	0 +8	3 +8	9 +8	2 +8	7 +8	1 +8	6 +8	4 +8	8 +8	5 +8
9.	1 +9	6 +9	4 +9	2 +9	8 +9	0 +9	5 +9	9 +9	3 +9	7 +9

WHOLE NUMBERS
Column Addition

In the exercises below, you will work with 3 or more addends. You can add down the column. You can check your answer by adding up the column.

Add down.   Add up.

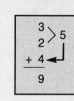

PRACTICE ───────────────────────────────────────

Add.

	a	b	c	d	e	f	g	h	i
1.	5	4	3	7	6	4	8	2	5
	4	4	6	2	3	5	1	7	3
	+ 3	+ 5	+ 2	+ 1	+ 6	+ 8	+ 3	+ 6	+ 8
	12								
2.	4	5	8	2	1	3	4	6	7
	9	7	1	7	4	5	3	5	5
	+ 6	+ 3	+ 4	+ 5	+ 8	+ 7	+ 6	+ 8	+ 6
3.	6	3	5	2	4	9	7	8	1
	2	4	3	7	3	2	3	1	9
	5	6	4	3	8	5	4	6	8
	+ 7	+ 5	+ 7	+ 8	+ 5	+ 7	+ 6	+ 7	+ 5
4.	1	7	3	2	4	6	8	5	7
	4	2	5	4	3	5	3	5	4
	9	4	7	8	6	8	5	7	3
	+ 7	+ 9	+ 6	+ 5	+ 8	+ 4	+ 6	+ 4	+ 9
5.	3	1	2	5	4	5	2	1	3
	2	6	4	3	2	5	3	6	3
	4	5	3	1	2	2	4	2	1
	7	3	6	4	9	5	1	5	4
	+ 8	+ 4	+ 5	+ 9	+ 5	+ 6	+ 7	+ 8	+ 9

Using a Calculator

The AC key on your calculator clears everything. The C key clears your last entry. Always clear your calculator when you begin a new problem. Press the AC key once or press the C key twice.

To add numbers use the + key and the = key. To find the sum of 9 and 7,

Enter 9. Press the + key.   (9)

Enter 7. Press the = key.   (16)

PRACTICE

Use your calculator to find the following sums. Remember: Always clear your calculator before you begin another problem.

	a	*b*	*c*	*d*
1.	9 + 5 = _14_	7 + 6 = ___	8 + 8 = ___	2 + 4 = ___
2.	3 + 5 = ___	4 + 9 = ___	6 + 0 = ___	1 + 8 = ___
3.	4 + 6 = ___	6 + 4 = ___	8 + 5 = ___	5 + 8 = ___
4.	7 + 9 = ___	9 + 7 = ___	3 + 6 = ___	6 + 3 = ___

Write the sums. Use your calculator to check your answers.

	a	*b*	*c*	*d*	*e*	*f*	*g*
5.	9 + 3	6 + 4	5 + 8	2 + 6	8 + 2	3 + 5	7 + 9
6.	0 + 9	9 + 0	4 + 8	8 + 4	6 + 5	5 + 6	0 + 0

Using a Calculator to Add More Than Two Numbers

To add more than two numbers, you use the $+$ more than once. To find the sum of 6 and 5 and 7,

Enter 6. Press the $+$ key.	6
Enter 5. Press the $+$ key.	11
Enter 7. Press the $=$ key.	18

PRACTICE ——

Use your calculator to find the following sums. Remember: Always clear your calculator before you begin another problem.

	a	b	c
1.	9 $+$ 5 $+$ 3 $=$ _____	2 $+$ 6 $+$ 4 $=$ _____	8 $+$ 3 $+$ 7 $=$ _____
2.	6 $+$ 8 $+$ 2 $=$ _____	8 $+$ 5 $+$ 6 $=$ _____	4 $+$ 9 $+$ 4 $=$ _____
3.	4 $+$ 3 $+$ 9 $=$ _____	9 $+$ 2 $+$ 7 $=$ _____	7 $+$ 5 $+$ 6 $=$ _____
4.	7 $+$ 9 $+$ 4 $=$ _____	4 $+$ 1 $+$ 8 $=$ _____	5 $+$ 2 $+$ 8 $=$ _____

Write the sums. Use your calculator to check your answers.

	a	b	c	d	e	f
5.	7 4 + 8	3 9 + 5	9 8 + 6	4 6 + 4	6 5 + 8	5 2 + 2
6.	4 2 5 + 3	2 9 1 + 8	7 5 8 + 5	5 6 0 + 7	3 3 3 + 3	8 3 5 + 9
7.	5 4 0 + 1	9 3 1 + 5	3 7 5 + 2	7 6 4 + 8	2 8 8 + 0	8 8 3 + 3

WHOLE NUMBERS

Reading and Writing Numbers

A place-value chart can help you understand whole numbers. Each digit in a number has a value based on its place in the number.

The 7 is in the millions place.
Its value is 7 millions or 7,000,000.

The 1 is in the ten thousands place.
Its value is 1 ten thousand or 10,000.

The 4 is in the hundreds place.
Its value is 4 hundreds or 400.

		hundred millions	ten millions	millions	hundred thousands	ten thousands	thousands	hundreds	tens	ones
			7	5	1	8	4	6	0	
			Millions			Thousands			Ones	

We read and write this number as: seven million, five hundred eighteen thousand, four hundred sixty.

Notice that commas are used to separate digits into groups of three. This helps make large numbers easier to read.

PRACTICE

Write the place name for the 4 in each number.

	a		*b*
1. 263,041	*tens*	3,340,687	
2. 921,426,050		4,531,906	

Write the value of the underlined digit.

	a		*b*
3. 3,2_5_7,461	*5 ten thousands*	54,69_9_	
4. _3_,257,461		54,_6_99	

Write each number using digits. Insert commas where needed.

5. six hundred forty-five thousand, three hundred ten _____ *645,310*

6. eighty-seven thousand, four hundred sixteen _____

7. ten thousand, eighty-nine _____

Write each number in words. Insert commas where needed.

8. 80,462 _____ *eighty thousand, four hundred sixty-two* _____

9. 506 _____

10. 12,934 _____

WHOLE NUMBERS
Comparing and Ordering

To compare two numbers, begin at the left.
Compare the digits in each place.

The symbol > means "is greater than." *5 > 3*
The symbol < means "is less than." *4 < 12*
The symbol = means "is equal to." *7 = 7*

Compare 42 and 27.

4 2
2 7   4 > 2 so,
       42 > 27

Compare 125 and 56.

1 2 5
0 5 6   1 > 0 so,
         125 > 56

Compare 324 and 374.

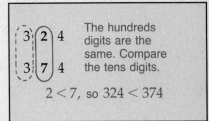

3 2 4   The hundreds
3 7 4   digits are the
         same. Compare
         the tens digits.

2 < 7, so 324 < 374

PRACTICE
Compare. Write >, <, or =.

a	b	c
1. 47 __<__ 73	86 _____ 70	73 _____ 200
2. 25 _____ 60	142 _____ 321	104 _____ 102
3. 50 _____ 50	304 _____ 107	408 _____ 1014
4. 1003 _____ 489	749 _____ 83	3030 _____ 3200
5. 607 _____ 6003	535 _____ 353	732 _____ 732
6. 7240 _____ 4188	3058 _____ 2907	5879 _____ 3004
7. 82,000 _____ 82,001	69,240 _____ 69,243	35,000 _____ 34,999

Write in order from least to greatest.

8. 12 34 26 ___ *12 26 34* ___

9. 425 643 523 _____

10. 247 617 338 _____

11. 133 873 245 _____

12. 71 170 107 _____

13. 836 335 683 _____

14. 647 538 134 _____

15. 384 916 700 _____

16. 4266 7109 3871 _____

WHOLE NUMBERS

Rounding

Rounded numbers tell about how many. You can use a number line to help you round numbers.

Remember, when a number is halfway, always round the number up.

Round 43 to the nearest ten.

40   43   50

43 is closer to 40 than to 50.

43 rounds down to 40.

Round 750 to the nearest hundred.

700   750   800

750 is halfway between 700 and 800.

750 rounds up to 800.

Round 3482 to the nearest hundred.

3400   3482   3500

3482 is closer to 3500 than to 3400.

3482 rounds up to 3500.

PRACTICE

Round to the nearest ten.

	a	b	c	d
1.	57 ___60___	82 _____	49 _____	35 _____
2.	34 _____	63 _____	76 _____	51 _____

Round to the nearest hundred.

	a	b	c	d
3.	471 ___500___	739 _____	850 _____	399 _____
4.	782 _____	456 _____	327 _____	218 _____
5.	139 _____	549 _____	744 _____	350 _____

Round to the nearest hundred.

	a	b	c	d
6.	2537 ___2500___	5499 _____	6205 _____	3668 _____
7.	8763 _____	6819 _____	3320 _____	8572 _____
8.	18,305 ___18,300___	44,174 _____	36,745 _____	11,831 _____

WHOLE NUMBERS

Adding Larger Numbers

Adding numbers with more than one digit is like doing column addition more than once. You start with the ones column.

Find: 722 + 263.

Add the ones.	Add the tens.	Add the hundreds.
H T O	H T O	H T O
7 2 2	7 2 2	7 2 2
+ 2 6 3	+ 2 6 3	+ 2 6 3
5	8 5	9 8 5

PRACTICE

Add.

	a	*b*	*c*	*d*	*e*
1.	633 +264 **897**	144 +650	337 +242	455 +324	382 + 17
2.	432 +345	720 +250	630 +307	274 +725	322 +475
3.	4536 +3343	3215 +3584	4167 +3522	3156 +5732	7135 +2564
4.	3516 +4102	1681 +4218	7006 +2893	8341 +1628	6328 +1050
5.	1275 +3521	2144 +7050	2369 +6530	4527 +4302	1733 +6254
6.	23,507 +14,280	32,404 +25,331	27,680 +41,206	40,582 +18,206	74,610 +25,389

Addition with Regrouping

When the sum in a column is more than 9, you must regroup that sum before adding the next column. Always start with the ones column.

Find: 538 + 354.

Add the ones. 8 + 4 = 12 Regroup.	Add the tens. 1 + 3 + 5 = 9	Add the hundreds. 5 + 3 = 8
H | T | O 　　1 5 | 3 | 8 +3 | 5 | 4 　　| 2	H | T | O 　　1 5 | 3 | 8 +3 | 5 | 4 　| 9 | 2	H | T | O 　　1 5 | 3 | 8 +3 | 5 | 4 8 | 9 | 2

PRACTICE

Add.

	a	*b*	*c*	*d*	*e*
1.	1 525 +356 881	338 +243	516 +247	429 +354	354 +237
2.	337 +427	212 +359	327 +555	235 +719	335 +528
3.	234 +439	424 +358	217 +374	249 +515	235 +518
4.	693 +134	223 +285	325 +193	581 +346	475 +281
5.	442 +285	573 +356	246 +193	457 +451	663 +274
6.	715 +156	392 +245	654 +173	527 +357	338 +116

More Addition with Regrouping

When the sum in a column is more than 9, you must regroup. Sometimes you must regroup more than once. Always start with the ones column.

Find: 648 + 573.

Add the ones. 8 + 3 = 11 Regroup.	Add the tens. 1 + 4 + 7 = 12 Regroup.	Add the hundreds. 1 + 6 + 5 = 12 Regroup.
Th | H | T | O 　　　　*1* 　　6 | 4 | 8 +　5 | 7 | 3 　　　　　1	Th | H | T | O 　　　*1* | *1* 　　6 | 4 | 8 +　5 | 7 | 3 　　　2 | 1	Th | H | T | O 　　*1* | 1 　　6 | 4 | 8 +　5 | 7 | 3 1 | 2 | 2 | 1

PRACTICE

Add.

	a	b	c	d	e
1.	*1 1* 247 +683 930	578 +243	490 +247	846 +374	764 +285
2.	234 +368	607 +409	583 + 75	35 +167	872 +500
3.	3234 + 439	6464 +1358	289 +7515	3005 +2897	15,225 + 7,951
4.	234 210 +352	157 60 +164	2395 2708 + 431	6394 2182 +2014	12,105 9,387 + 1,532

Line up the digits. Then add.

	a	b	c
5.	731 + 294 = _____ 731 +294	3755 + 2881 = _____	7357 + 2983 = _____
6.	5312 + 748 = _____	3009 + 7896 = _____	876 + 5499 = _____

WHOLE NUMBERS
Estimation of Sums

To estimate a sum, first round each number to the same place. Then add the rounded numbers.

Estimate: 856 + 431

Round each number to the same place. Add.

$$856 \rightarrow 900$$
$$+431 \rightarrow + 400$$
$$\overline{ 1300}$$

Estimate: 1583 + 632

Round each number to the same place. Add.

$$1583 \rightarrow 1600$$
$$+ 632 \rightarrow + 600$$
$$\overline{ 2200}$$

PRACTICE

Estimate the sum.

	a	b	c	d
1.	$327 \rightarrow 300$ $+492 \rightarrow +500$ $\overline{800}$	$253 \rightarrow$ $+485 \rightarrow$	$384 \rightarrow$ $+234 \rightarrow$	$265 \rightarrow$ $+341 \rightarrow$
2.	$728 \rightarrow$ $+371 \rightarrow$	$452 \rightarrow$ $+285 \rightarrow$	$167 \rightarrow$ $+462 \rightarrow$	$272 \rightarrow$ $+594 \rightarrow$
3.	$543 \rightarrow$ $+488 \rightarrow$	$754 \rightarrow$ $+376 \rightarrow$	$486 \rightarrow$ $+540 \rightarrow$	$535 \rightarrow$ $+480 \rightarrow$
4.	$924 \rightarrow$ $+668 \rightarrow$	$822 \rightarrow$ $+555 \rightarrow$	$733 \rightarrow$ $+248 \rightarrow$	$612 \rightarrow$ $+145 \rightarrow$
5.	$1287 \rightarrow$ $+ 739 \rightarrow$	$2105 \rightarrow$ $+ 618 \rightarrow$	$1692 \rightarrow$ $+1487 \rightarrow$	$3492 \rightarrow$ $+ 877 \rightarrow$

Addition Practice With and Without Regrouping

PRACTICE

Add.

	a	*b*	*c*	*d*	*e*
1.	*1 1* 657 +267 *9 2 4*	768 + 94	987 +346	779 + 83	693 +835
2.	418 +793	846 +457	575 +666	704 +599	236 +421
3.	684 +203	796 +709	985 + 57	394 +505	756 +986
4.	6779 +1834	6945 +2243	4967 +6021	3512 +3894	3852 +1948
5.	4397 +8056	3947 + 874	5693 +1876	3207 +2984	8006 +3995
6.	3805 +4778	6784 +8522	2411 +6228	3528 +2497	2896 +1105

Line up the digits. Then add.

	a	*b*	*c*
7.	$651 + 398 =$ ___ 651 + 398	$436 + 253 =$ ___	$6138 + 446 =$ ___
8.	$2968 + 1885 =$ ___	$4008 + 2995 =$ ___	$4321 + 2859 =$ ___
9.	$7562 + 2314 =$ ___	$6384 + 2645 =$ ___	$1234 + 8967 =$ ___

WHOLE NUMBERS

More Addition Practice With and Without Regrouping

PRACTICE

Add.

	a	b	c	d	e
1.	569 +769	379 +824	753 +234	716 +761	695 +346
2.	125 +878	675 +204	346 +263	629 +152	736 +162
3.	421 +535	413 +897	591 +257	637 +906	590 +845
4.	2798 + 425	4207 + 634	9325 + 987	6724 + 595	1388 + 208
5.	3805 +4778	6784 +8522	2411 +6729	3528 +2417	2896 +9107
6.	3469 +4578	1046 +5997	2964 +7092	2180 +6609	9619 +3624
7.	23 51 +13	76 63 +98	836 979 + 13	2703 144 + 32	854 825 +945
8.	528 788 + 47	807 605 +796	73 93 +1856	378 245 +876	3667 79 + 974

WHOLE NUMBERS

Addition Practice Fun

Solve the number clues.
Use the sums to complete the number puzzle.

Across

1. $57 + 24 = \_$

3. $89 + 7 = \_$

5. $2 + 1 + 3 + 2 = \_$

6. $1638 + 2301 = \_$

9. $574 + 245 = \_$

11. $2 + 4 + 0 + 3 = \_$

12. $3690 + 7059 = \_$

14. $34 + 48 = \_$

15. $764 + 457 = \_$

17. $43 + 8 = \_$

18. $385 + 437 = \_$

19. $4 + 3 + 2 = \_$

20. $25 + 57 = \_$

21. $38 + 41 = \_$

22. $26 + 45 = \_$

23. $7239 + 4585 = \_$

25. $6 + 1 + 2 = \_$

26. $639 + 284 = \_$

27. $76 + 16 = \_$

Down

1. $562 + 277 = \_$

2. $9 + 2 + 8 = \_$

3. $2 + 3 + 4 = \_$

4. $4899 + 63,523 = \_$

5. $54 + 35 = \_$

7. $187 + 125 = \_$

8. $68 + 22 = \_$

10. $109 + 83 = \_$

13. $583,269 + 129,674 = \_$

14. $3547 + 4574 = \_$

16. $9897 + 9295 = \_$

17. $234 + 347 = \_$

18. $5204 + 3518 = \_$

22. $3 + 0 + 1 + 3 = \_$

24. $23 + 66 = \_$

27. $2 + 1 + 4 + 2 = \_$

WHOLE NUMBERS

The Subtraction Facts

Shown below are subtraction facts that you need to know very well. Practice subtracting until you have the facts memorized.

PRACTICE

Subtract.

	a	b	c	d	e	f	g	h	i	j
1.	6 −1	10 − 6	4 − 0	6 − 5	10 − 1	11 − 7	7 − 4	2 − 0	11 − 6	3 − 0
2.	12 − 3	12 − 8	12 − 5	3 − 1	14 − 5	8 − 1	4 − 4	10 − 5	9 − 7	9 − 8
3.	14 − 7	7 − 5	13 − 6	11 − 8	8 − 6	10 − 4	9 − 6	11 − 9	7 − 3	12 − 4
4.	4 − 1	7 − 7	11 − 2	11 − 3	15 − 8	7 − 6	13 − 9	8 − 7	8 − 4	9 − 3
5.	9 − 2	17 − 9	13 − 4	10 − 3	14 − 6	17 − 8	11 − 5	16 − 8	16 − 7	11 − 4
6.	8 − 2	9 − 5	15 − 7	13 − 7	18 − 9	7 − 2	5 − 3	12 − 9	6 − 4	4 − 3
7.	15 − 6	16 − 9	15 − 9	10 − 7	5 − 1	10 − 9	12 − 6	6 − 2	8 − 5	13 − 5
8.	14 − 8	3 − 3	8 − 3	9 − 9	8 − 8	6 − 3	2 − 2	12 − 4	14 − 9	12 − 7
9.	13 − 8	10 − 8	5 − 5	2 − 1	5 − 2	9 − 4	5 − 4	9 − 1	4 − 2	10 − 2

Subtracting Numbers

Subtracting numbers with two or three digits is like subtracting basic facts
two or three times. You start with the ones column.

Find: 168 − 23.

Subtract the ones.	Subtract the tens.	Subtract the hundreds.
H T O 1 6 8 − 2 3 5	H T O 1 6 8 − 2 3 4 5	H T O 1 6 8 − 2 3 1 4 5

PRACTICE ———————————————————————————

Subtract.

	a	b	c	d	e
1.	65 −52 13	89 −46	56 −46	98 −37	62 −40
2.	83 −21	76 −12	84 − 3	98 −15	56 − 5
3.	57 −33	98 −50	78 −13	97 −42	82 −21
4.	89 −62	68 −51	52 −42	57 − 7	67 −16
5.	123 − 11	596 − 62	247 − 14	498 − 87	166 − 40
6.	239 − 17	153 − 21	419 − 16	367 − 32	384 − 80

WHOLE NUMBERS

Subtracting Larger Numbers

Subtracting numbers with two or more digits is like subtracting basic facts again and again. Always start with the ones column.

Find: 5864 − 2143.

Subtract the ones.	Subtract the tens.	Subtract the hundreds.	Subtract the thousands.

	Th	H	T	O
	5	8	6	4
−	2	1	4	3
				1

	Th	H	T	O
	5	8	6	4
−	2	1	4	3
			2	1

	Th	H	T	O
	5	8	6	4
−	2	1	4	3
		7	2	1

	Th	H	T	O
	5	8	6	4
−	2	1	4	3
	3	7	2	1

PRACTICE

Subtract.

	a	b	c	d	e
1.	989 −756 = 233	346 −236	637 −124	875 −425	798 −350
2.	495 −130	752 −742	769 −452	836 −305	528 −412
3.	5398 − 175	6839 − 407	1765 − 245	3567 − 416	1998 − 362
4.	2286 − 215	1475 − 322	3854 − 701	1445 − 215	2690 − 580
5.	1584 −1423	3975 −1772	5391 −4131	2879 −1679	4155 −3015
6.	3875 −1562	2215 −1100	6489 −4050	5768 −1233	1995 −1985

28

WHOLE NUMBERS

Subtracting with Regrouping

Sometimes when you try to subtract in the ones column, you do not have enough to subtract. You must regroup a ten to get enough ones.

Find: 32 − 15.

Can you subtract the ones? No.		Regroup.		Subtract the ones. $12 - 5 = 7$		Subtract the tens. $2 - 1 = 1$	
Tens	Ones	Tens	Ones	Tens	Ones	Tens	Ones
3	2	$\overset{2}{\cancel{3}}$	$\overset{12}{\cancel{2}}$	$\overset{2}{\cancel{3}}$	$\overset{12}{\cancel{2}}$	$\overset{2}{\cancel{3}}$	$\overset{12}{\cancel{2}}$
− 1	5	− 1	5	− 1	5	− 1	5
					7	1	7

PRACTICE

Subtract.

	a	b	c	d	e
1.	$\overset{4\,10}{\cancel{5\,0}}$ − 1 2 3 8	8 1 − 3 7	6 6 − 3 9	8 1 − 7	5 0 − 1 8
2.	8 1 − 2 9	3 4 − 8	2 6 − 9	6 3 − 1 7	5 7 − 9
3.	8 6 − 3 9	7 6 − 3 8	7 5 − 2 8	9 6 − 3 7	4 5 − 6
4.	4 2 − 1 3	3 8 − 1 9	8 6 − 3 8	6 5 − 3 9	8 2 − 5

If there are enough ones to subtract, then do so. If not, regroup first.

	a	b	c	d	e
5.	5 7 − 1 5	6 0 − 2 3	5 6 − 2 4	9 2 − 2 1	6 4 − 2 9
6.	9 1 − 4 3	6 7 − 2 5	9 1 − 4 2	5 7 − 2 5	8 6 − 5

Subtracting with Regrouping Hundreds

Sometimes you can subtract in the ones column, but you do not have enough tens to subtract. Then you must regroup hundreds as 1 less hundred and 10 more tens.

Find: 329 − 146.

Subtract the ones. 9 − 6 = 3	Regroup the tens.	Subtract the tens. 12 − 4 = 8	Subtract the hundreds. 2 − 1 = 1

H T O
3 2 9
− 1 4 6
_____
    3
	H T O
	²3̷ ¹²2̷ 9
	− 1 4 6
	_____
	    3
		H T O
		²3̷ ¹²2̷ 9
		− 1 4 6
		_____
		  8 3
			H T O
			²3̷ ¹²2̷ 9
			− 1 4 6
			_____
			1 8 3

PRACTICE

Subtract.

	a	b	c	d	e
1.	⁵¹⁶ 6̷6̷8 − 3 9 1 _____ 2 7 7	4 4 9 − 8 6	8 3 3 − 1 7 2	5 5 1 − 4 7 0	3 0 6 − 1 5 2
2.	9 8 9 − 1 9 7	9 1 9 − 7 2 2	3 6 0 − 2 8 0	5 5 6 − 6 1	6 4 8 − 1 6 2
3.	8 4 8 − 2 9 0	8 7 8 − 6 9 7	4 3 1 − 1 5 1	8 7 9 − 1 8 0	2 2 2 − 9 0
4.	3 2 9 − 6 3	6 0 8 − 5 8 5	4 3 8 − 2 8 8	8 1 9 − 3 3 8	5 5 8 − 2 8 6

If there are enough to subtract, then do so. If not, regroup first.

	a	b	c	d	e
5.	9 2 2 − 7 3 2	7 3 9 − 5 2 4	9 0 2 − 1 2 1	4 2 3 − 2 1 8	8 6 8 − 5 2 3
6.	6 0 8 − 5 3	4 7 2 − 3 8 1	9 7 6 − 2 1 5	8 5 9 − 3 8 4	5 1 1 − 3 7 0

Subtracting with Regrouping Hundreds

When you subtract, sometimes you do not need to regroup at all. Sometimes you must regroup once. Sometimes you must regroup more than once.

Find: 320 − 158.

Regroup 1 ten as 10 ones. Subtract. 10 − 8 = 2	Regroup 1 hundred as 10 tens. Subtract. 11 − 5 = 6	Subtract the hundreds. 2 − 1 = 1
$\begin{array}{c|c|c} H & T & O \\ & 1 & 10 \\ 3 & \cancel{2} & \cancel{0} \\ -1 & 5 & 8 \\ \hline & & 2 \end{array}$	$\begin{array}{c|c|c} H & T & O \\ & 11 & \\ 2 & \cancel{1} & 10 \\ \cancel{3} & \cancel{2} & \cancel{0} \\ -1 & 5 & 8 \\ \hline & 6 & 2 \end{array}$	$\begin{array}{c|c|c} H & T & O \\ & 11 & \\ 2 & \cancel{1} & 10 \\ \cancel{3} & \cancel{2} & \cancel{0} \\ -1 & 5 & 8 \\ \hline 1 & 6 & 2 \end{array}$

PRACTICE

Subtract.

	a	*b*	*c*	*d*	*e*
1.	$\begin{array}{r} {\scriptstyle 9} \\ {\scriptstyle 5\ 10\ 10} \\ \cancel{600} \\ -241 \\ \hline 359 \end{array}$	$\begin{array}{r} 357 \\ -169 \\ \hline \end{array}$	$\begin{array}{r} 518 \\ -329 \\ \hline \end{array}$	$\begin{array}{r} 734 \\ -468 \\ \hline \end{array}$	$\begin{array}{r} 952 \\ -687 \\ \hline \end{array}$
2.	$\begin{array}{r} 946 \\ -847 \\ \hline \end{array}$	$\begin{array}{r} 542 \\ -275 \\ \hline \end{array}$	$\begin{array}{r} 516 \\ -189 \\ \hline \end{array}$	$\begin{array}{r} 734 \\ -\ 87 \\ \hline \end{array}$	$\begin{array}{r} 342 \\ -256 \\ \hline \end{array}$
3.	$\begin{array}{r} 561 \\ -478 \\ \hline \end{array}$	$\begin{array}{r} 787 \\ -389 \\ \hline \end{array}$	$\begin{array}{r} 635 \\ -\ 57 \\ \hline \end{array}$	$\begin{array}{r} 532 \\ -346 \\ \hline \end{array}$	$\begin{array}{r} 438 \\ -359 \\ \hline \end{array}$

Subtract. You may have to regroup twice, once, or not at all.

4.	$\begin{array}{r} 632 \\ -159 \\ \hline \end{array}$	$\begin{array}{r} 434 \\ -189 \\ \hline \end{array}$	$\begin{array}{r} 531 \\ -128 \\ \hline \end{array}$	$\begin{array}{r} 462 \\ -189 \\ \hline \end{array}$	$\begin{array}{r} 435 \\ -198 \\ \hline \end{array}$
5.	$\begin{array}{r} 902 \\ -301 \\ \hline \end{array}$	$\begin{array}{r} 388 \\ -119 \\ \hline \end{array}$	$\begin{array}{r} 650 \\ -178 \\ \hline \end{array}$	$\begin{array}{r} 507 \\ -229 \\ \hline \end{array}$	$\begin{array}{r} 800 \\ -356 \\ \hline \end{array}$

WHOLE NUMBERS
Estimation of Differences

To estimate a difference, first round each number to the same place. Then subtract the rounded numbers.

Estimate: 654 − 210

Round each number to the same place.
Subtract.

$$
\begin{array}{r}
654 \rightarrow 700 \\
-210 \rightarrow -200 \\
\hline
500
\end{array}
$$

Estimate: 3794 − 832

Round each number to the same place.
Subtract.

$$
\begin{array}{r}
3794 \rightarrow 3800 \\
-832 \rightarrow -800 \\
\hline
3000
\end{array}
$$

PRACTICE

Estimate the difference.

	a	b	c	d

1.
$$
\begin{array}{r}
482 \rightarrow 500 \\
-246 \rightarrow -200 \\
\hline
300
\end{array}
\qquad
\begin{array}{r}
357 \rightarrow \\
-129 \rightarrow \\
\hline
\end{array}
\qquad
\begin{array}{r}
568 \rightarrow \\
-374 \rightarrow \\
\hline
\end{array}
\qquad
\begin{array}{r}
845 \rightarrow \\
-659 \rightarrow \\
\hline
\end{array}
$$

2.
$$
\begin{array}{r}
682 \rightarrow \\
-249 \rightarrow \\
\hline
\end{array}
\qquad
\begin{array}{r}
453 \rightarrow \\
-326 \rightarrow \\
\hline
\end{array}
\qquad
\begin{array}{r}
376 \rightarrow \\
-165 \rightarrow \\
\hline
\end{array}
\qquad
\begin{array}{r}
928 \rightarrow \\
-572 \rightarrow \\
\hline
\end{array}
$$

3.
$$
\begin{array}{r}
546 \rightarrow \\
-366 \rightarrow \\
\hline
\end{array}
\qquad
\begin{array}{r}
765 \rightarrow \\
-249 \rightarrow \\
\hline
\end{array}
\qquad
\begin{array}{r}
430 \rightarrow \\
-357 \rightarrow \\
\hline
\end{array}
\qquad
\begin{array}{r}
828 \rightarrow \\
-680 \rightarrow \\
\hline
\end{array}
$$

4.
$$
\begin{array}{r}
1098 \rightarrow \\
-726 \rightarrow \\
\hline
\end{array}
\qquad
\begin{array}{r}
1473 \rightarrow \\
-815 \rightarrow \\
\hline
\end{array}
\qquad
\begin{array}{r}
2591 \rightarrow \\
-866 \rightarrow \\
\hline
\end{array}
\qquad
\begin{array}{r}
2283 \rightarrow \\
-599 \rightarrow \\
\hline
\end{array}
$$

5.
$$
\begin{array}{r}
3388 \rightarrow \\
-1357 \rightarrow \\
\hline
\end{array}
\qquad
\begin{array}{r}
4042 \rightarrow \\
-2279 \rightarrow \\
\hline
\end{array}
\qquad
\begin{array}{r}
5910 \rightarrow \\
-1873 \rightarrow \\
\hline
\end{array}
\qquad
\begin{array}{r}
6491 \rightarrow \\
-4185 \rightarrow \\
\hline
\end{array}
$$

WHOLE NUMBERS

Addition and Subtraction: Mixed Practice

PRACTICE

Add or subtract. Watch the signs that tell you which operation to use.

	a	b	c	d	e
1.	96 +21	48 +37	514 +263	817 +146	753 +289
2.	57 −12	82 −27	658 −324	592 −257	800 −536
3.	6142 +3756	9625 +4213	2358 +4135	6834 +8879	5705 + 986
4.	8654 −2431	9642 −6425	7194 −3457	9426 −3258	9000 −5427
5.	582 143 +214	861 495 +827	779 827 +579	926 79 +658	4214 3942 +3463
6.	38,762 −26,521	45,962 −12,535	37,424 −15,258	76,200 −41,350	90,000 −12,542
7.	27,454 +41,345	75,633 +83,698	946 +30,306	35,878 +89,646	92,027 +21,973

Line up the digits. Then add or subtract.

a	b
8. 9787 + 1578 + 463 + 9242 = _____	750,000 − 7500 = _____

PROBLEM-SOLVING STRATEGY

Write a Number Sentence

Sometimes it helps to write a number sentence to solve a problem. Read the problem and look for the action that is involved. This will help you know which operation to use in the number sentence.

If two or more parts, things, or groups are put together or joined, then use addition.

If a whole is separated into parts, or if parts are being compared, then use subtraction.

EXAMPLE 1
Read the problem.

On the opening day of their new restaurant, Alex and Alexis served 528 customers at lunch time and 842 customers at dinner time. How many customers did they serve on opening day?

Solve the problem.

Join or put together the customers to find how many in all. Write and solve an addition sentence.

$$528 + 842 = n$$
$$528 + 842 = 1370$$

They served 1370 customers.

EXAMPLE 2
Read the problem.

On the second day, Alex and Alexis served 1872 customers in all. If they served 743 customers at lunch time, how many customers did they serve at dinner time?

Solve the problem.

Separate the whole into two parts. You know one part. To find the other part, write and solve a subtraction sentence.

$$1872 - 743 = n$$
$$1872 - 743 = 1129$$

They served 1129 customers at dinner time.

Write and solve a number sentence.

1. Alex and Alexis served 1370 customers on the first day and 1872 on the second day. How many more customers did they serve on the second day than they did on the first?

   $1872 - 1370 = n$

   $1872 - 1370 = 502$

   Answer _____

2. Alex and Alexis say that they will feel successful if they can serve 8400 customers in their first week. After the first two days, how many more customers must they serve on the next 5 days to reach their goal? [Hint: Use the numbers in Problem 1 to help you find the answer.]

   Answer _____

3. A food supplier who serves Alex and Alexis has 25 cases of fresh lettuce in one warehouse, 84 cases in a second warehouse, and 42 cases in a third warehouse. How many cases of lettuce does the supplier have in all?

   Answer _____

4. The food supplier buys 90 crates of fresh melon from a farmer. Twenty-seven crates are loaded onto the supplier's first truck to arrive at the farm. How many more crates remain to be loaded onto trucks?

   Answer _____

5. Alex estimates that the restaurant will use about 8500 paper napkins each week. Just to be on the safe side, he decides to have on hand another 500 napkins. How many napkins will Alex order for one week?

   Answer _____

6. Alexis orders the containers for carry-out food. She makes sure that she has 6500 containers at the beginning of each week. If she has 320 containers left over from last week, how many containers will she order for the new week?

   Answer _____

7. The restaurant is on the border of New Hampshire and Vermont. The population of New Hampshire is 920,610. Vermont has 511,456 people. Alex wishes that all of them come to the restaurant some day. How many people is he wishing will come?

   Answer _____

8. Although she now lives in New Hampshire, Alexis grew up along the Great Lakes. What is the difference between the areas of Lake Superior and Lake Erie? Lake Superior covers 81,000 square miles, and Lake Erie covers 32,630 square miles.

   Answer _____

Applications

Solve.

1. Marie budgeted the following for her monthly bills: $450 for rent and $100 for electricity. How much did she budget altogether for these bills?

Answer _____

2. The Jackson family traveled 4234 miles during vacation one year and 5378 miles the next year. How many miles did they travel during the two vacations?

Answer _____

3. The area of North Carolina is 48,843 square miles. South Carolina's area is 30,207 square miles. Which is larger? How much larger?

Answer _____

4. What is the total area of North and South Carolina? (See Problem 3.)

Answer _____

5. Pluto has a diameter of 5600 kilometers. Jupiter has a diameter of 142,000 kilometers. How much larger than Pluto is Jupiter?

Answer _____

6. Neptune's year has 60,195 days and Uranus's year has 30,685 days. How much longer is Neptune's year than Uranus's year?

Answer _____

7. In 1990, the population of Arkansas was 2,350,725. Arizona had 3,665,228 people. Which state had more people? How many more?

Answer _____

8. What was the total population of Arkansas and Arizona in 1990? (See problem 7.)

Answer _____

9. The attendance at the Raider's football game one week was 5479. The next week 4388 people attended the game. Altogether, how many people attended these two games?

Answer _____

10. Mount Whitney is 14,494 feet high. Borah Peak is 12,662 feet high. How much higher than Borah Peak is Mount Whitney?

Answer _____

WHOLE NUMBERS

Unit 1 Review

Estimate the answers.

	a	b	c	d
1.	$289 \rightarrow$ $+126 \rightarrow$	$7163 \rightarrow$ $+1615 \rightarrow$	$716 \rightarrow$ $-225 \rightarrow$	$6819 \rightarrow$ $-2196 \rightarrow$

Add.

	a	b	c	d	e
2.	72 $+\ 5$	48 $+\ 9$	62 $+16$	58 $+26$	325 $+164$
3.	618 $+129$	542 $+376$	494 $+258$	709 $+395$	3275 $+6948$

Subtract.

	a	b	c	d	e
4.	29 $-\ 6$	68 -23	45 $-\ 9$	83 -26	124 $-\ 68$
5.	735 -212	652 -346	419 -125	823 -265	3900 $-\ 439$

Solve.

6. A jet liner flew 760 miles from Atlanta, Georgia, to New York City. Then it flew 749 miles to Chicago, Illinois. How many miles did it fly in all?

Answer _____

7. Two jet liners took off from Los Angeles, California. One flew 2611 miles to Boston, Massachusetts and the other flew 1246 miles to Dallas, Texas. How much farther did the Boston plane fly than the Dallas plane?

Answer _____

WHOLE NUMBERS

The Multiplication Table

×	0	1	2	3	4	5	6	7	8	9
0	0	0	0	0	0	0	0	0	0	0
1	0	1	2	3	4	5	6	7	8	9
2	0	2	4	6	8	10	12	14	16	18
3	0	3	6	9	12	15	18	21	24	27
4	0	4	8	12	16	20	24	28	32	36
5	0	5	10	15	20	25	30	35	40	45
6	0	6	12	18	24	30	36	42	48	54
7	0	7	14	21	28	35	42	49	56	63
8	0	8	16	24	32	40	48	56	64	72
9	0	9	18	27	36	45	54	63	72	81

The table above shows the multiplication facts. To find 7×6, move across the 7 row and down the 6 column. Where they meet is their product, 42.

PRACTICE

Use the table to answer the following questions.

1. Each number in the 7 row is how much more than the number before it?

_____

2. Each number in the 8 row is how much more than the number before it?

_____

3. Each number in the 3 row is how much more than the number before it?

_____

4. Each number in the 5 row is how much more than the number before it?

_____

5. Each number in the 7 column is how much more than the number above it?

_____

6. Each number in the 8 column is how much more than the number above it?

_____

7. Each number in the 2 column is how much more than the number above it?

_____

8. Each number in the 6 column is how much more than the number above it?

_____

WHOLE NUMBERS

The Multiplication Facts

Shown below are some multiplication facts that you need to know very well. Practice multiplying until you have the facts memorized.

PRACTICE

Multiply.

	a	b	c	d	e	f	g	h	i	j
1.	5 ×1	4 ×6	7 ×0	1 ×5	9 ×1	4 ×7	3 ×4	2 ×0	5 ×6	3 ×0
2.	9 ×3	4 ×8	7 ×5	2 ×1	9 ×5	7 ×1	0 ×4	5 ×5	2 ×7	1 ×8
3.	7 ×7	2 ×5	7 ×6	3 ×8	2 ×6	6 ×4	3 ×6	2 ×9	4 ×3	8 ×4
4.	3 ×1	0 ×7	9 ×2	8 ×3	7 ×8	1 ×6	4 ×9	1 ×7	4 ×4	6 ×3
5.	7 ×2	8 ×9	9 ×4	7 ×3	8 ×6	9 ×8	6 ×5	8 ×8	9 ×7	7 ×4
6.	6 ×2	4 ×5	8 ×7	6 ×7	9 ×9	5 ×2	2 ×3	3 ×9	2 ×4	1 ×3
7.	9 ×6	7 ×9	6 ×9	3 ×7	4 ×1	1 ×9	6 ×6	4 ×2	3 ×5	8 ×5
8.	6 ×8	0 ×3	5 ×3	0 ×9	8 ×0	3 ×3	0 ×2	8 ×4	5 ×9	5 ×7
9.	5 ×8	2 ×8	1 ×5	1 ×1	3 ×2	5 ×4	1 ×4	8 ×1	2 ×2	8 ×2

Multiplying Larger Numbers

When multiplying larger numbers, use basic multiplication facts more than once. Always multiply the ones digits first.

Find: 2 × 34.

Multiply the ones.	Multiply the tens.
2 × 4 ones = 8 ones	2 × 3 tens = 6 tens

	H	T	O
		3	4
×			2
			8

	H	T	O
		3	4
×			2
		6	8

Find: 12 × 13.

Multiply ones by ones.	Multiply ones by tens.	Multiply tens by ones.	Multiply tens by tens.
2 × 3 ones = 6 ones	2 × 1 ten = 2 tens	Use 0 as a placeholder. 1 ten × 3 = 3 tens	1 ten × 1 ten = 1 hundred Add the partial products.

	H	T	O
		1	3
×		1	2
			6

	H	T	O
		1	3
×		1	2
		2	6

	H	T	O
		1	3
×		1	2
		2	6
		3	0

	H	T	O
		1	3
×		1	2
		2	6
	1	3	0
	1	5	6

PRACTICE

Multiply.

	a	b	c	d	e
1.	12 × 4 48	32 × 3	21 × 7	52 × 4	24 × 2
2.	12 × 12	23 × 13	42 × 21	61 × 11	24 × 22
3.	16 × 11	21 × 41	11 × 11	33 × 22	32 × 23

Multiplying with Regrouping

Sometimes it is necessary to regroup a product when multiplying larger numbers. The regrouping is similar to what you did in addition.

Find: 4 × 58.

Find 13 × 29.

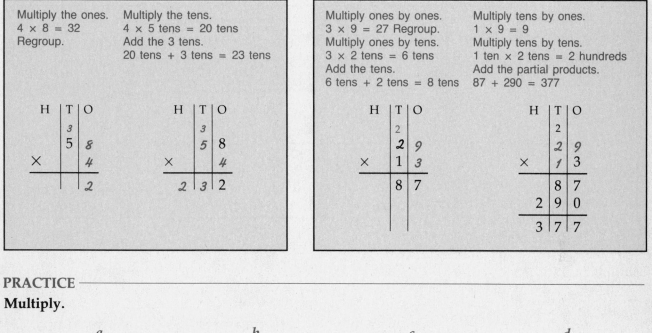

PRACTICE ——————————————————————

Multiply.

1.

	a		b		c		d	
	H T O		H T O		H T O		H T O	

a: 3 5 9 × 4 = 2 3 6

b: 2 8 × 3

c: 2 7 × 1 3

d: 3 4 × 2 3

	a	b	c	d	e
2.	27 × 6	25 × 4	42 × 9	36 × 3	17 × 8
3.	152 × 3	252 × 3	224 × 3	325 × 3	116 × 5
4.	24 × 32	18 × 12	13 × 43	31 × 21	59 × 17

WHOLE NUMBERS

More Multiplying with Regrouping

Sometimes it is necessary to regroup more than once when multiplying.
Always start multiplying with the ones digits.

Find: 76 × 368.

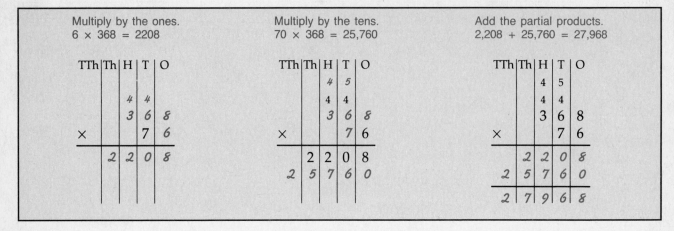

	Multiply by the ones. 6 × 368 = 2208	Multiply by the tens. 70 × 368 = 25,760	Add the partial products. 2,208 + 25,760 = 27,968

PRACTICE

Multiply.

1.

	a		b		c		d
TTh Th H T O		TTh Th H T O		TTh Th H T O		TTh Th H T O	

a.
```
    2
    5
    4 7
  × 3 8
    3 7 6
  1 4 1 0
  1 7 8 6
```

b.
```
  3 5 6 4
  ×   2 4
```

c.
```
  1 4 6 2
  ×   5 2
```

d.
```
  4 3 7 8
  ×   1 3
```

	a	b	c	d	e
2.	57 ×38	245 × 4	436 × 29	238 × 49	537 × 8
3.	86 ×64	209 × 35	784 × 9	98 ×75	514 × 46
4.	304 × 95	273 × 28	511 × 50	128 ×264	539 ×645

42

Multiply.

	a	b	c	d	e
1.	168 × 5	459 × 7	45 ×32	80 ×60	24 ×38
2.	764 × 46	300 × 23	453 × 50	851 × 78	352 × 24
3.	532 ×290	497 ×308	936 ×500	354 ×655	307 ×465

Line up the digits. Then multiply.

a	b	c
4. 318 × 7 = _____	936 × 4 = _____	521 × 9 = _____

$$\begin{array}{r} 318 \\ \times\ \ 7 \\ \hline \end{array}$$

a	b	c
5. 474 × 86 = _____	231 × 32 = _____	566 × 27 = _____

▶ MIXED PRACTICE

Find each answer.

	a	b	c	d	e
1.	231 +709	7684 − 251	8957 +2082	5306 −2859	2270 − 183
2.	57,236 −29,810	81,244 +17,038	41,000 − 7,455	83,201 +49,839	14,835 + 790

Multiplication Practice

PRACTICE

Multiply.

	a	*b*	*c*	*d*	*e*
1.	231 × 3	412 × 2	112 × 4	332 × 3	212 × 4
2.	769 × 8	825 × 6	892 × 8	483 × 9	536 × 7
3.	849 × 8	675 × 9	935 × 5	684 × 7	324 × 8
4.	18 × 70	56 × 20	49 × 30	37 × 50	25 × 10
5.	53 × 25	46 × 47	90 × 58	38 × 57	28 × 16
6.	456 × 13	308 × 18	764 × 46	354 × 55	418 × 23
7.	597 × 28	936 × 37	284 × 163	635 × 207	712 × 328

The Division Facts

If you know the multiplication facts, mastering the division facts should be easy. For example, 28 ÷ 7 asks, "How many 7's are in 28?" Since you know that 4 × 7 = 28, then you know that the answer, or quotient, is 4.

PRACTICE

Complete the sentences.

	a	b
1.	Since 3 × 9 = 27, then 27 ÷ 9 = _____	Since 6 × 4 = 24, then 24 ÷ 4 = _____
2.	Since 2 × 8 = 16, then 8$\overline{)16}$ = _____	Since 5 × 9 = 45, then 9$\overline{)45}$ = _____

Divide.

	a	b	c	d	e	f	g	h	i	j
3.	$1\overline{)5}$	$6\overline{)24}$	$7\overline{)0}$	$5\overline{)5}$	$1\overline{)9}$	$7\overline{)28}$	$4\overline{)12}$	$5\overline{)0}$	$6\overline{)30}$	$3\overline{)3}$
4.	$3\overline{)27}$	$8\overline{)32}$	$5\overline{)35}$	$1\overline{)2}$	$5\overline{)45}$	$7\overline{)7}$	$4\overline{)0}$	$5\overline{)25}$	$7\overline{)14}$	$8\overline{)8}$
5.	$7\overline{)49}$	$5\overline{)10}$	$6\overline{)42}$	$8\overline{)24}$	$6\overline{)12}$	$4\overline{)24}$	$6\overline{)18}$	$9\overline{)18}$	$3\overline{)12}$	$4\overline{)32}$
6.	$1\overline{)3}$	$8\overline{)0}$	$2\overline{)18}$	$3\overline{)24}$	$8\overline{)56}$	$6\overline{)6}$	$9\overline{)36}$	$1\overline{)6}$	$4\overline{)16}$	$3\overline{)18}$
7.	$2\overline{)14}$	$9\overline{)72}$	$4\overline{)36}$	$3\overline{)21}$	$6\overline{)48}$	$8\overline{)72}$	$5\overline{)30}$	$8\overline{)64}$	$7\overline{)63}$	$4\overline{)28}$
8.	$2\overline{)12}$	$5\overline{)20}$	$7\overline{)56}$	$7\overline{)42}$	$9\overline{)81}$	$2\overline{)10}$	$3\overline{)6}$	$9\overline{)27}$	$4\overline{)8}$	$6\overline{)0}$
9.	$6\overline{)54}$	$9\overline{)63}$	$9\overline{)54}$	$7\overline{)21}$	$1\overline{)4}$	$9\overline{)9}$	$6\overline{)36}$	$2\overline{)8}$	$5\overline{)15}$	$5\overline{)40}$
10.	$8\overline{)48}$	$3\overline{)0}$	$3\overline{)15}$	$9\overline{)0}$	$1\overline{)7}$	$3\overline{)9}$	$2\overline{)2}$	$1\overline{)0}$	$9\overline{)45}$	$7\overline{)35}$
11.	$8\overline{)40}$	$8\overline{)16}$	$2\overline{)0}$	$1\overline{)1}$	$2\overline{)6}$	$4\overline{)20}$	$4\overline{)4}$	$1\overline{)8}$	$2\overline{)4}$	$2\overline{)16}$

Division with 1-digit Divisors

To divide by a 1-digit divisor, first decide on a trial quotient. Then multiply and subtract. Write the remainders in the quotient.

Remember, if your trial quotient is too large or too small, try another number.

Find: 430 ÷ 4

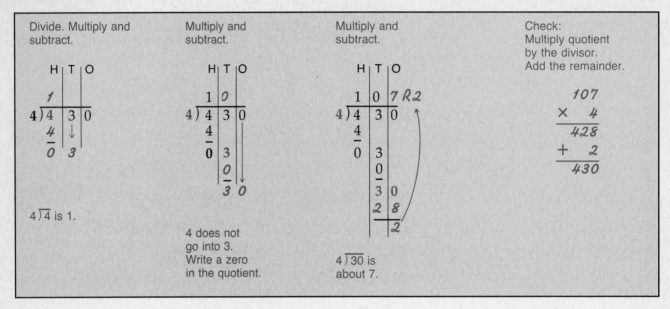

GUIDED PRACTICE

Divide. Check.

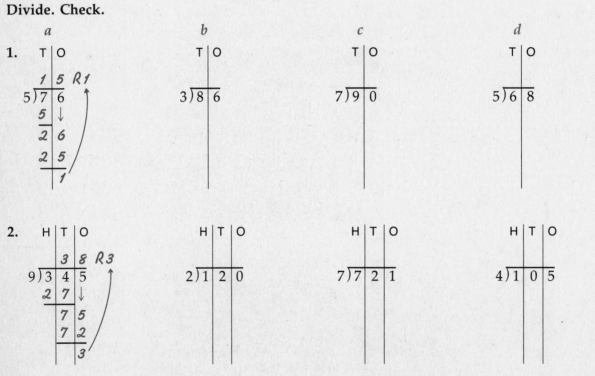

Divide.

	a	b	c	d	e
1.	3)69	4)87	5)54	9)92	6)59
2.	4)568	6)726	7)800	3)434	9)308
3.	5)148	8)160	8)253	4)743	3)534
4.	6)100	9)907	5)619	7)440	4)652

Set up the problem. Then divide.

	a	b	c
5.	128 ÷ 2 = _____	57 ÷ 9 = _____	283 ÷ 4 = _____
	2)128		

Find each answer.

	a	b	c	d
1.	381 295 + 32	6183 −1766	43,925 + 9,014	30,000 −10,921
2.	375 × 8	107 × 9	451 × 32	876 × 57

Division Practice

Here are two hints to help you know if your trial quotient is correct.

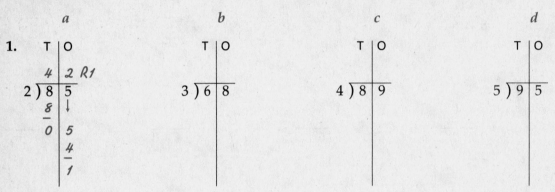

If the remainder is more than the divisor, then your trial quotient is not enough.

```
        5    ← Not        Just →      6
    4)264     enough.     right.   4)264
     20                             24
      6                              2
```

If you cannot subtract after multiplying the trial quotient and the divisor, then your trial quotient is too much.

```
        7    ← Too        Just →      6
    4)276     much.       right.   4)276
     28                             24
                                     3
```

GUIDED PRACTICE

Divide. Check.

a   b   c   d

1.

```
   T O
   4|2 R1
2)8 5
  8|↓
  0|5
    4
    1
```

	T	O
3)	6	8

	T	O
4)	8	9

	T	O
5)	9	5

2.

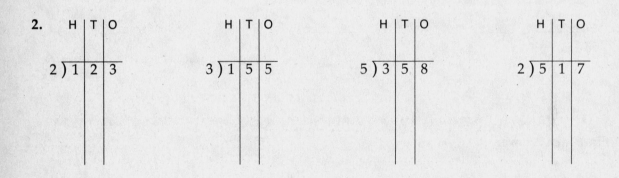

	H	T	O
2)	1	2	3

	H	T	O
3)	1	5	5

	H	T	O
5)	3	5	8

	H	T	O
2)	5	1	7

3.

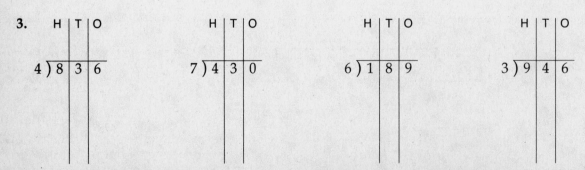

	H	T	O
4)	8	3	6

	H	T	O
7)	4	3	0

	H	T	O
6)	1	8	9

	H	T	O
3)	9	4	6

Divide. Check.

	a	b	c	d	e
1.	$3\overline{)64}$	$4\overline{)86}$	$7\overline{)94}$	$5\overline{)75}$	$2\overline{)87}$

	a	b	c	d	e
2.	$5\overline{)259}$	$7\overline{)225}$	$3\overline{)146}$	$9\overline{)188}$	$6\overline{)433}$

	a	b	c	d	e
3.	$2\overline{)384}$	$8\overline{)977}$	$4\overline{)915}$	$7\overline{)699}$	$3\overline{)814}$

	a	b	c	d	e
4.	$6\overline{)618}$	$3\overline{)205}$	$5\overline{)102}$	$9\overline{)918}$	$4\overline{)796}$

Set up the problem. Then divide.

	a	b	c
5.	$627 \div 2 = $ _____	$459 \div 5 = $ _____	$721 \div 3 = $ _____

MIXED PRACTICE
Find each answer.

	a	b	c	d
1.	$\begin{array}{r} 495 \\ 208 \\ +\ \ 77 \\ \hline \end{array}$	$\begin{array}{r} 6021 \\ -2913 \\ \hline \end{array}$	$\begin{array}{r} 8709 \\ +4058 \\ \hline \end{array}$	$\begin{array}{r} 2800 \\ -1765 \\ \hline \end{array}$
2.	$\begin{array}{r} 559 \\ \times\ \ \ 6 \\ \hline \end{array}$	$\begin{array}{r} 604 \\ \times\ \ \ 5 \\ \hline \end{array}$	$\begin{array}{r} 725 \\ \times\ \ 17 \\ \hline \end{array}$	$\begin{array}{r} 408 \\ \times\ \ 37 \\ \hline \end{array}$

49

WHOLE NUMBERS

Division with 2-digit Divisors

To divide by a 2-digit divisor, decide on a trial quotient. Multiply and subtract. Write the remainder in the quotient.

Find: 748 ÷ 35

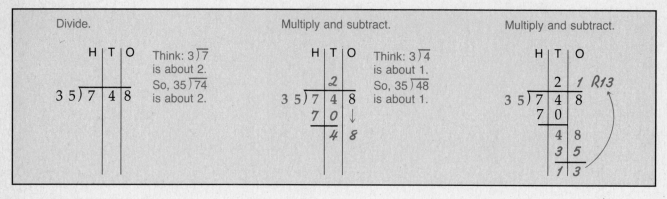

GUIDED PRACTICE

Divide.

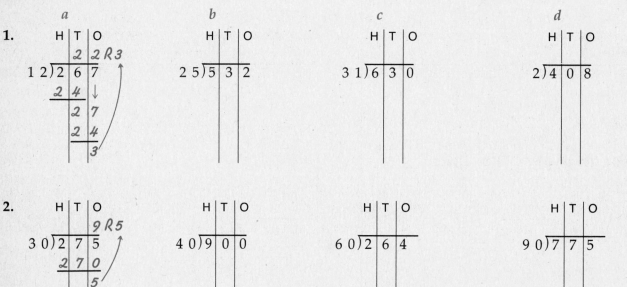

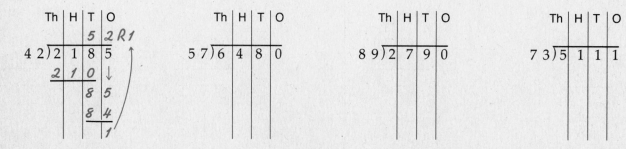

Divide.

	a	b	c	d
1.	36)733	40)982	89)700	26)793
2.	81)923	72)841	32)980	47)562
3.	45)2430	30)2914	63)4536	59)3827

Set up the problem. Then divide.

	a	b	c
4.	840 ÷ 73 = _____	761 ÷ 51 = _____	694 ÷ 33 = _____

73)840

➡ MIXED PRACTICE
Find each answer.

	a	b	c	d
1.	324 +299	8375 −1981	238 × 7	905 × 6
2.	423 × 21	806 − 95	252 × 44	766 + 83

Division Practice

Sometimes when you divide, your quotient may not be great enough or it may be too great. Here is how you can tell.

<table>
<tr><td>

If the remainder is more than the divisor, then your quotient is not great enough.

```
        7  ← Not        Just →        8  R12
  46)380     enough.   right. 46)380
  322                          368
   58                           12
```

</td><td>

If you cannot subtract after multiplying the quotient and the divisor, then your quotient is too great.

```
        7  ← Too        Just →        6  R27
  33)225     great.    right. 33)225
  231                          198
                                27
```

</td></tr>
</table>

PRACTICE

Divide.

	a	b	c	d
1.	$\begin{array}{r} 3\ R27 \\ 73\overline{)246} \\ 219 \\ \hline 27 \end{array}$	$47\overline{)230}$	$33\overline{)175}$	$47\overline{)240}$
2.	$76\overline{)541}$	$52\overline{)412}$	$41\overline{)283}$	$28\overline{)194}$
3.	$84\overline{)331}$	$31\overline{)245}$	$57\overline{)291}$	$36\overline{)220}$
4.	$18\overline{)132}$	$38\overline{)274}$	$32\overline{)288}$	$89\overline{)582}$
5.	$92\overline{)550}$	$27\overline{)171}$	$69\overline{)356}$	$26\overline{)162}$

Dividing Larger Numbers

To divide greater numbers by a 2-digit divisor, decide on a trial quotient.
Multiply and subtract. Check to make sure the quotient is great enough
and not too great. Write any remainder after the quotient.

Find: 96,446 ÷ 26.

Divide. Multiply. Subtract.	Divide. Multiply. Subtract.	Divide. Multiply. Subtract.	Divide. Multiply. Subtract.
$$\begin{array}{r} 3 \\ 26\overline{)96,446} \\ 78\!\downarrow \\ \hline 184 \end{array}$$	$$\begin{array}{r} 3,7 \\ 26\overline{)96,446} \\ 78 \\ \hline 184 \\ 182\!\downarrow \\ \hline 24 \end{array}$$	$$\begin{array}{r} 3,7\,0 \\ 26\overline{)96,446} \\ 78 \\ \hline 184 \\ 182 \\ \hline 24\!\downarrow \\ 0\!\downarrow \\ \hline 246 \end{array}$$	$$\begin{array}{r} 3,7\,0\,9 \text{ R12} \\ 26\overline{)96,446} \\ 78 \\ \hline 184 \\ 182 \\ \hline 24 \\ 0 \\ \hline 246 \\ 234 \\ \hline 12 \end{array}$$

PRACTICE

Divide.

	a	*b*	*c*	*d*
1.	$$\begin{array}{r} 314 \\ 75\overline{)23,550} \\ 225 \\ \hline 105 \\ 75 \\ \hline 300 \\ 300 \\ \hline 0 \end{array}$$	$42\overline{)54,364}$	$62\overline{)60,202}$	$89\overline{)95,720}$
2.	$92\overline{)77,777}$	$28\overline{)14,448}$	$39\overline{)67,612}$	$59\overline{)10,830}$

WHOLE NUMBERS

Dividing by Hundreds

When you divide by hundreds, look at the first 3 or 4 digits of the dividend to determine your first trial quotient. From that point on, use the same steps: divide, multiply, and subtract.

Find: 52,386 ÷ 388.

Divide.
Multiply. Subtract.

```
        1
388)52,386
    388↓
    135 8
```

Divide.
Multiply. Subtract.

```
       13
388)52,386
    388 |
    1358 |
    1164↓
    194 6
```

Divide.
Multiply. Subtract.

```
      135 R6
388)52,386
    388
    1358
    1164
    1946
    1940
       6
```

PRACTICE

Divide.

	a	b	c	d

1.

```
      233 R107
341)79,560
    682
    1136
    1023
     1130
     1023
      107
```

b. 453)18,327

c. 236)3317

d. 426)3966

2.

863)63,043

904)75,000

506)435,160

246)123,456

WHOLE NUMBERS

Multiplication and Division Practice

Multiply.

	a	b	c	d
1.	72 × 4	319 × 3	1728 × 9	50,800 × 4
2.	37 × 24	78 × 56	485 × 92	6948 × 89
3.	9967 × 36	45,847 × 65	592 × 231	6342 × 358

Divide.

	a	b	c	d
4.	63)834	9)8452	4)2563	7)36,533
5.	24)96	27)5776	14)1032	26)8174
6.	85)71,995	60)28,988	57)34,676	357)44,982

WHOLE NUMBERS

Estimation of Products

To estimate products, round each number. Then multiply the rounded numbers.

Estimate: 37 × 54

Round each number to the same place. Multiply.

$$
\begin{array}{r}
37 \rightarrow \quad 40 \\
\times 54 \rightarrow \times 50 \\
\hline
2000
\end{array}
$$

Estimate: 25 × 211

Round each number to the same place. Multiply.

$$
\begin{array}{r}
211 \rightarrow \quad 210 \\
\times \; 25 \rightarrow \times \; 30 \\
\hline
6300
\end{array}
$$

PRACTICE

Estimate the product.

a	b	c	d
1. $\begin{array}{r} 43 \rightarrow \quad 40 \\ \times 82 \rightarrow \times 80 \\ \hline 3200 \end{array}$	$\begin{array}{r} 27 \rightarrow \\ \times 14 \rightarrow \\ \hline \end{array}$	$\begin{array}{r} 58 \rightarrow \\ \times 33 \rightarrow \\ \hline \end{array}$	$\begin{array}{r} 76 \rightarrow \\ \times 65 \rightarrow \\ \hline \end{array}$
2. $\begin{array}{r} 83 \rightarrow \\ \times 26 \rightarrow \\ \hline \end{array}$	$\begin{array}{r} 77 \rightarrow \\ \times 55 \rightarrow \\ \hline \end{array}$	$\begin{array}{r} 49 \rightarrow \\ \times 37 \rightarrow \\ \hline \end{array}$	$\begin{array}{r} 31 \rightarrow \\ \times 94 \rightarrow \\ \hline \end{array}$
3. $\begin{array}{r} 193 \rightarrow \\ \times \; 41 \rightarrow \\ \hline \end{array}$	$\begin{array}{r} 788 \rightarrow \\ \times \; 29 \rightarrow \\ \hline \end{array}$	$\begin{array}{r} 572 \rightarrow \\ \times \; 15 \rightarrow \\ \hline \end{array}$	$\begin{array}{r} 299 \rightarrow \\ \times \; 38 \rightarrow \\ \hline \end{array}$
4. $\begin{array}{r} 746 \rightarrow \quad 700 \\ \times 526 \rightarrow \times 500 \\ \hline 350,000 \end{array}$	$\begin{array}{r} 892 \rightarrow \\ \times 175 \rightarrow \\ \hline \end{array}$	$\begin{array}{r} 632 \rightarrow \\ \times 362 \rightarrow \\ \hline \end{array}$	$\begin{array}{r} 719 \rightarrow \\ \times 428 \rightarrow \\ \hline \end{array}$

Estimate the product.

a	b	c
5. 38 × 66 $\begin{array}{r} 66 \rightarrow \\ \times 38 \rightarrow \\ \hline \end{array}$	945 × 13	211 × 747
6. 756 × 912	61 × 183	22 × 59

Estimation of Quotients

To estimate quotients, use rounded numbers.

Estimate: 424 ÷ 6

Round to use a basic fact.	Divide.

$6\overline{)424}$ Think: 6 × 7 = 42 $6\overline{)420}^{70}$

Estimate: 928 ÷ 29

Round each number.	Divide.

928 → 900
29 → 30 $30\overline{)900}^{30}$

PRACTICE

Estimate using basic facts.

a	*b*	*c*

1. $4\overline{)362} \rightarrow 4\overline{)360}^{90}$ $7\overline{)558}$ $8\overline{)404}$

2. $5\overline{)3005} \rightarrow 5\overline{)3000}^{600}$ $9\overline{)7222}$ $6\overline{)2432}$

Estimate by rounding both numbers.

a	*b*	*c*

3. $26\overline{)598} \rightarrow 30\overline{)600}^{20}$ $38\overline{)812}$ $19\overline{)589}$

4. $12\overline{)664}$ $34\overline{)648}$ $42\overline{)630}$

5. $61\overline{)632}$ $52\overline{)936}$ $54\overline{)756}$

PROBLEM-SOLVING STRATEGY
Choose an Operation

Sometimes a problem does not tell you whether to add or subtract, multiply or divide. To solve such a problem, you must read the problem carefully. Then, decide what the problem is asking you to do. Next, choose an operation and solve the problem.

EXAMPLE 1
Read the problem.

> The three heaviest players on a football team weigh 290 pounds, 278 pounds, and 304 pounds. What is the total weight of these three players?

Decide what the problem is asking.

> In this problem, the question "What is the total . . .?" is asking you to find a sum.

Choose the operation.

> To solve, you must add.

Solve the problem.

> $290 + 278 + 304 = 872$
>
> The total weight of the three players is 872 pounds.

EXAMPLE 2
Read the problem.

> Wong works 38 hours a week. How many hours will he work in 4 weeks?

Decide what the problem is asking.

> In this problem the question "How many . . .?" is asking for a total.

Choose the operation.

> To solve, you must multiply 38 by 4. Of course you could add 38 four times, but adding would take longer.

Solve the problem.

> $38 \times 4 = 152$
>
> Wong will work 152 hours in 4 weeks.

Choose the correct operation needed to solve each problem. Then solve the problem.

1. The distance from Chicago, Illinois to Butte, Montana is 1522 miles. Seattle, Washington is 567 miles beyond Butte. How far is it from Chicago to Seattle?

   Operation _____

   Answer _____

2. A department store sold 6 television sets for a total of $3120. Each set sold for the same price. What was the price of each television set?

   Operation _____

   Answer _____

3. A freight train has 70 cars. Each car can hold 50,000 pounds of material. How much weight can the train hold in all?

   Operation _____

   Answer _____

4. Jerome counted 845 old books and 519 new books for the book sale. How many more old books were sold than new books?

   Operation _____

   Answer _____

5. Fletcher deposited a check for $292. He received $105 back in cash. What was the total amount of his deposit?

   Operation _____

   Answer _____

6. A boxcar can carry 1540 bushels of wheat. A bushel of wheat weighs 60 pounds. How many pounds of wheat can a boxcar carry?

   Operation _____

   Answer _____

7. Sue scored a 78, an 81, a 76, and a 79 in the golf tournament. What was her total score?

   Operation _____

   Answer _____

8. On their vacation, the Carlins drove an average of 55 miles per hour for 10 hours. How many miles did they drive on their vacation?

   Operation _____

   Answer _____

PROBLEM SOLVING

Applications

Solve.

1. Each car on a commuter train seats 54 passengers. The train has 8 cars. How many people can be seated on the train?

Answer _____

2. Louisa spends $10 on lunch for each 5-day school week. She spends the same amount each day. How much does Louisa spend for each lunch?

Answer _____

3. Miami is 665 miles from Atlanta. Rita's car gets 35 miles per gallon of gasoline. How much gasoline will she use to drive from Miami to Atlanta?

Answer _____

4. The land area in Alaska is 570,833 square miles. Hawaii is 6427 square miles. How much larger is Alaska than Hawaii?

Answer _____

5. Lawrence jogs 3 miles a day. How many miles does he jog in a week? How many miles does he jog in 4 weeks?

Answer _____

6. Peter's yearly salary is $24,000. How much does Peter earn each month?

Answer _____

7. The highest point in Florida is 345 feet above sea level. New Jersey's highest point is 1803 feet above sea level. How much higher than Florida's highest point is New Jersey's highest point?

Answer _____

8. It is 710 miles from the northern border of Texas to its southern border. The state is 760 miles from east to west. How much farther is it between the east/west border than between the north/south border?

Answer _____

WHOLE NUMBERS

Unit 2 Review

Estimate the answers.

	a	b	c	d

1. $37 \rightarrow$ $48 \rightarrow$ $5\overline{)463} \rightarrow$ $39\overline{)1645} \rightarrow$
    $\times 12 \rightarrow$ $\times 39 \rightarrow$

Multiply.

	a	b	c	d

2. $\begin{array}{r} 425 \\ \times\ \ \ 6 \end{array}$ $\begin{array}{r} 2307 \\ \times\ \ \ \ \ 4 \end{array}$ $\begin{array}{r} 62 \\ \times 23 \end{array}$ $\begin{array}{r} 58 \\ \times 46 \end{array}$

3. $\begin{array}{r} 674 \\ \times\ \ 20 \end{array}$ $\begin{array}{r} 908 \\ \times\ \ 63 \end{array}$ $\begin{array}{r} 2587 \\ \times\ \ \ \ 49 \end{array}$ $\begin{array}{r} 452 \\ \times 378 \end{array}$

Divide.

	a	b	c	d

4. $5\overline{)865}$ $8\overline{)3572}$ $3\overline{)12{,}654}$ $42\overline{)139}$

5. $32\overline{)7455}$ $48\overline{)5624}$ $40\overline{)92{,}571}$ $62\overline{)19{,}576}$

Solve.

6. During their vacation, the Andersen family drove an average of 324 miles a day. If their vacation was 7 days long, how many miles in all did they drive?

7. During their vacation, Mr. and Mrs. Cortez flew on an international jet liner that traveled 6375 miles in 15 hours. What was the jet's average speed per hour?

Answer _____

Answer _____

THE MEANING AND USE OF FRACTIONS

Proper Fractions

A fraction names part of a whole. This circle has four equal parts. Each part is $\frac{1}{4}$ of the circle.

1 of the 4 equal parts is shaded blue.

numerator ↘
$\frac{1}{4}$ — one blue part
— four parts in all
denominator ↗

We read $\frac{1}{4}$ as one fourth.

A fraction also names part of a group. Three of the four triangles are shaded blue.

$\frac{3}{4}$ — three blue
— four in all

Three fourths are blue.

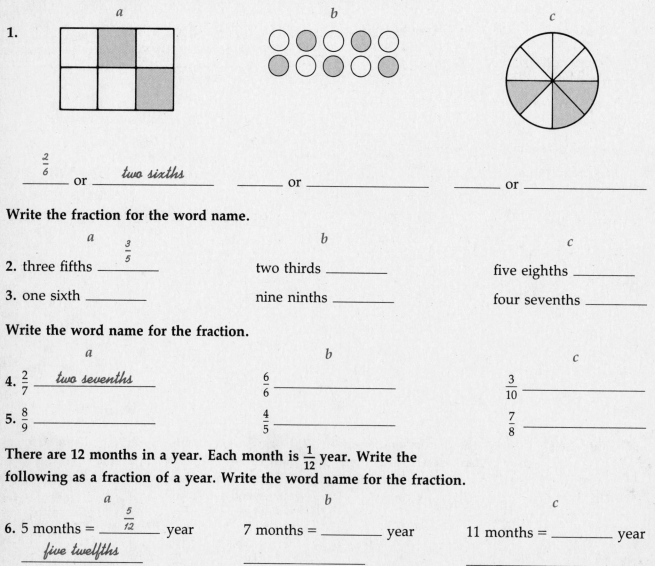

PRACTICE

Write the fraction and the word name for the part that is shaded.

         a   b   c

1.

$\frac{2}{6}$
_____ or ___ *two sixths* ___   _____ or _____   _____ or _____

Write the fraction for the word name.

        a   b   c

2. three fifths ___ $\frac{3}{5}$ ___   two thirds _____   five eighths _____

3. one sixth _____   nine ninths _____   four sevenths _____

Write the word name for the fraction.

        a   b   c

4. $\frac{2}{7}$ ___ *two sevenths* ___ $\frac{6}{6}$ _____ $\frac{3}{10}$ _____

5. $\frac{8}{9}$ _____ $\frac{4}{5}$ _____ $\frac{7}{8}$ _____

There are 12 months in a year. Each month is $\frac{1}{12}$ year. Write the following as a fraction of a year. Write the word name for the fraction.

        a   b   c

6. 5 months = ___ $\frac{5}{12}$ ___ year   7 months = _____ year   11 months = _____ year
      five twelfths

THE MEANING AND USE OF FRACTIONS

Improper Fractions and Mixed Numbers

An improper fraction is a fraction with a numerator that is greater than or equal to the denominator.

$\frac{5}{5}$, $\frac{15}{5}$ and $\frac{13}{5}$ are improper fractions.

An improper fraction can be written as a whole or mixed number.

Write $\frac{5}{5}$ and $\frac{15}{5}$ as whole numbers.

Write $\frac{13}{5}$ as a mixed number.

A mixed number is a whole number and a fraction.

$1\frac{5}{12}$ is a mixed number.

A mixed number can be written as an improper fraction.

Write $1\frac{5}{12}$ as an improper fraction.

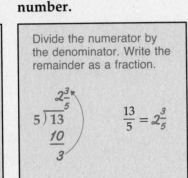

Divide the numerator by the denominator.

$5\overline{)5}^{1}$ $\frac{5}{5} = 1$

$5\overline{)15}^{3}$ $\frac{15}{5} = 3$

Divide the numerator by the denominator. Write the remainder as a fraction.

$5\overline{)13}^{2\frac{3}{5}}$ $\frac{13}{5} = 2\frac{3}{5}$
$\underline{10}$
3

Multiply the whole number and the denominator. Add this product to the numerator. Then write the sum over the denominator.

$1\frac{5}{12} = \frac{1 \times 12 + 5}{12} = \frac{12 + 5}{12} = \frac{17}{12}$

So, $1\frac{5}{12} = \frac{17}{12}$.

PRACTICE

Write as a mixed number or whole number.

	a	b	c	d
1.	$\frac{24}{6} = $ ___ 4 ___	$\frac{27}{9} = $ _____	$\frac{30}{15} = $ _____	$\frac{12}{12} = $ _____
2.	$\frac{13}{12} = $ ___ $1\frac{1}{12}$ ___	$\frac{3}{2} = $ _____	$\frac{4}{3} = $ _____	$\frac{5}{4} = $ _____
3.	$\frac{11}{8} = $ _____	$\frac{36}{6} = $ _____	$\frac{7}{4} = $ _____	$\frac{9}{3} = $ _____
4.	$\frac{64}{4} = $ _____	$\frac{11}{5} = $ _____	$\frac{16}{8} = $ _____	$\frac{25}{6} = $ _____

Write as an improper fraction.

	a	b	c	d
5.	$4\frac{1}{2} = $ ___ $\frac{9}{2}$ ___	$5\frac{4}{5} = $ _____	$6\frac{2}{3} = $ _____	$7\frac{1}{4} = $ _____
6.	$2\frac{7}{10} = $ _____	$8\frac{2}{9} = $ _____	$17\frac{3}{5} = $ _____	$9\frac{5}{8} = $ _____

Equivalent Fractions

To add or subtract fractions, you might need to change a fraction to an equivalent form. To change a fraction to an equivalent fraction in higher terms, multiply the numerator and the denominator by the same nonzero number.

Rewrite $\frac{2}{3}$ with 12 as the denominator.

Compare the denominators.	Multiply both the numerator and the denominator by 4.
$\frac{2}{3} = \frac{}{12}$ Think: $3 \times 4 = 12$	$\frac{2}{3} = \frac{2 \times 4}{3 \times 4} = \frac{8}{12}$

To add or subtract fractions, you might need to find the lowest common denominator (LCD) of the fractions. Then you can use the LCD to write equivalent fractions.

Use the LCD to write equivalent fractions for $\frac{2}{3}$ and $\frac{1}{4}$.

List several multiples of each denominator.	Find the LCD. It is the smallest number that appears on both lists.	Write equivalent fractions.
Multiples of 3: 3 6 9 12 15 Multiples of 4: 4 8 12 16 20	The LCD of $\frac{2}{3}$ and $\frac{1}{4}$ is 12.	$\frac{2}{3} = \frac{2 \times 4}{3 \times 4} = \frac{8}{12}$ $\frac{1}{4} = \frac{1 \times 3}{4 \times 3} = \frac{3}{12}$

PRACTICE

Rewrite each fraction as an equivalent fraction in higher terms.

	a	b	c	d
1.	$\frac{2}{3} = \frac{2 \times 3}{3 \times 3} = \frac{6}{9}$	$\frac{1}{2} = \frac{}{30}$	$\frac{3}{4} = \frac{}{12}$	$\frac{2}{5} = \frac{}{15}$
2.	$\frac{1}{2} = \frac{}{14}$	$\frac{5}{6} = \frac{}{18}$	$\frac{4}{5} = \frac{}{10}$	$\frac{3}{8} = \frac{}{16}$

Use the LCD to write equivalent fractions.

	a	b	c	d
3.	$\frac{3}{4} = \frac{3 \times 3}{4 \times 3} = \frac{9}{12}$	$\frac{1}{3} =$	$\frac{2}{5} =$	$\frac{5}{6} =$
	$\frac{1}{6} = \frac{1 \times 2}{6 \times 2} = \frac{2}{12}$	$\frac{4}{5} =$	$\frac{1}{4} =$	$\frac{7}{8} =$
4.	$\frac{2}{3} =$	$\frac{9}{10} =$	$\frac{1}{2} =$	$\frac{1}{3} =$
	$\frac{5}{7} =$	$\frac{3}{4} =$	$\frac{3}{5} =$	$\frac{1}{2} =$

Simplifying Fractions

When you find the answer to a problem with fractions, you might need to change the fraction to an equivalent form in simplest terms. To simplify a fraction, divide both the numerator and the denominator by the greatest number possible.

Simplify: $\frac{8}{10}$

Consider the numerator and denominator.

$\frac{8}{10} =$ Think: 10 can be divided by 5 but 8 cannot.
8 can be divided by 4 but 10 cannot.
Both 10 and 8 can be divided by 2.

Divide the numerator and the denominator by 2.

$\frac{8}{10} = \frac{8 \div 2}{10 \div 2} = \frac{4}{5}$

A fraction is in simplest terms when 1 is the only number that divides both the numerator and denominator evenly. The fraction $\frac{4}{5}$ is in simplest terms.

PRACTICE

Simplify.

	a	b	c	d
1.	$\frac{6}{8} = \frac{3}{4}$	$\frac{10}{20} =$	$\frac{3}{9} =$	$\frac{9}{12} =$
2.	$\frac{10}{12} =$	$\frac{8}{20} =$	$\frac{2}{8} =$	$\frac{4}{6} =$
3.	$\frac{5}{15} =$	$\frac{14}{16} =$	$\frac{8}{10} =$	$\frac{10}{25} =$
4.	$\frac{2}{4} =$	$\frac{6}{9} =$	$\frac{2}{6} =$	$\frac{3}{15} =$
5.	$\frac{12}{18} =$	$\frac{4}{12} =$	$\frac{2}{10} =$	$\frac{9}{21} =$
6.	$\frac{8}{14} =$	$\frac{2}{12} =$	$\frac{9}{15} =$	$\frac{15}{45} =$

THE MEANING AND USE OF FRACTIONS

Addition and Subtraction of Fractions with Like Denominators

To add or subtract fractions with like denominators, add or subtract the numerators. Use the same denominator. Simplify the answer.

Remember,
- to simplify an improper fraction, write it as a whole number or mixed number.
- to simplify a proper fraction, write it in simplest terms.

Find: $\frac{2}{7} + \frac{5}{7}$

Find: $\frac{7}{8} - \frac{3}{8}$

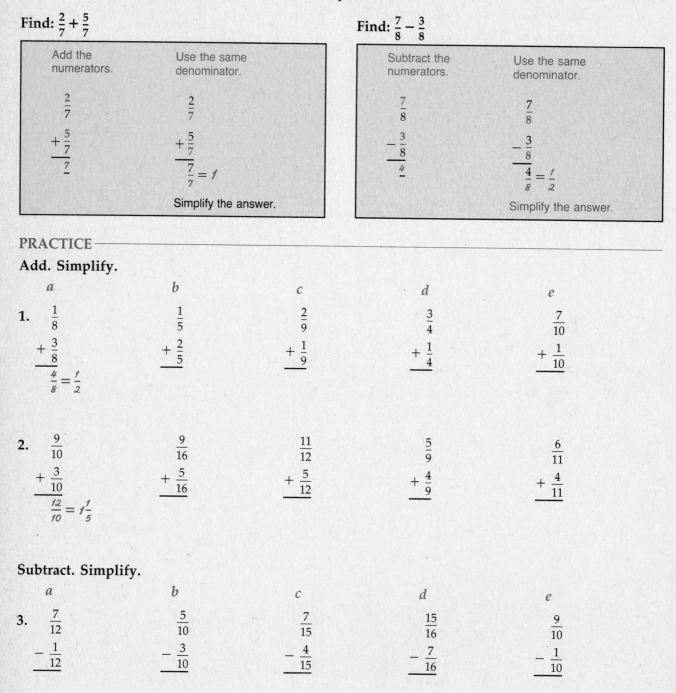

Add the numerators.	Use the same denominator.
$\frac{2}{7}$	$\frac{2}{7}$
$+\frac{5}{7}$	$+\frac{5}{7}$
$\frac{}{7}$	$\frac{7}{7} = 1$
	Simplify the answer.

Subtract the numerators.	Use the same denominator.
$\frac{7}{8}$	$\frac{7}{8}$
$-\frac{3}{8}$	$-\frac{3}{8}$
$\frac{4}{}$	$\frac{4}{8} = \frac{1}{2}$
	Simplify the answer.

PRACTICE

Add. Simplify.

	a	b	c	d	e
1.	$\frac{1}{8}$ $+\frac{3}{8}$ $\frac{4}{8} = \frac{1}{2}$	$\frac{1}{5}$ $+\frac{2}{5}$	$\frac{2}{9}$ $+\frac{1}{9}$	$\frac{3}{4}$ $+\frac{1}{4}$	$\frac{7}{10}$ $+\frac{1}{10}$
2.	$\frac{9}{10}$ $+\frac{3}{10}$ $\frac{12}{10} = 1\frac{1}{5}$	$\frac{9}{16}$ $+\frac{5}{16}$	$\frac{11}{12}$ $+\frac{5}{12}$	$\frac{5}{9}$ $+\frac{4}{9}$	$\frac{6}{11}$ $+\frac{4}{11}$

Subtract. Simplify.

	a	b	c	d	e
3.	$\frac{7}{12}$ $-\frac{1}{12}$	$\frac{5}{10}$ $-\frac{3}{10}$	$\frac{7}{15}$ $-\frac{4}{15}$	$\frac{15}{16}$ $-\frac{7}{16}$	$\frac{9}{10}$ $-\frac{1}{10}$

THE MEANING AND USE OF FRACTIONS

Addition of Fractions with Different Denominators

To add fractions with different denominators, first rewrite the fractions as equivalent fractions with like denominators. Then add the numerators and simplify the answer.

Find: $\frac{1}{4} + \frac{8}{16}$

Write equivalent fractions with like denominators.	Add the numerators. Use the same denominator.
$\frac{1}{4} = \frac{4}{16}$ Remember, $+\frac{8}{16} = \frac{8}{16}$ $\frac{1}{4} = \frac{1 \times 4}{4 \times 4} = \frac{4}{16}$	$\frac{1}{4} = \frac{4}{16}$ $+\frac{8}{16} = \frac{8}{16}$ $\frac{12}{16} = \frac{3}{4}$ Simplify the answer.

PRACTICE

Add. Simplify.

	a	*b*	*c*	*d*
1.	$\frac{5}{6} = \frac{5}{6}$ $+\frac{1}{2} = \frac{3}{6}$ $\frac{8}{6} = 1\frac{1}{3}$	$\frac{3}{8}$ $+\frac{1}{2}$	$\frac{1}{8}$ $+\frac{3}{4}$	$\frac{5}{8}$ $+\frac{1}{2}$
2.	$\frac{7}{9}$ $+\frac{2}{3}$	$\frac{5}{9}$ $+\frac{1}{3}$	$\frac{3}{10}$ $+\frac{1}{5}$	$\frac{2}{3}$ $+\frac{4}{21}$
3.	$\frac{2}{3}$ $+\frac{7}{15}$	$\frac{2}{5}$ $+\frac{9}{15}$	$\frac{3}{7}$ $+\frac{5}{14}$	$\frac{1}{20}$ $+\frac{7}{10}$

Add. Simplify.

	a	*b*	*c*
4.	$\frac{7}{10} + \frac{2}{5} =$ _____	$\frac{9}{16} + \frac{7}{8} =$ _____	$\frac{2}{21} + \frac{4}{7} =$ _____

$\frac{7}{10}$

$+\frac{2}{5}$

THE MEANING AND USE OF FRACTIONS

Addition of Fractions with Different Denominators Using the LCD

Find: $\frac{2}{3} + \frac{1}{5}$

Write equivalent fractions with like denominators. Use the LCD.

$$\frac{2}{3} = \frac{2 \times 5}{3 \times 5} = \frac{10}{15}$$

$$+\frac{1}{5} = \frac{1 \times 3}{5 \times 3} = \frac{3}{15}$$

Add the numerators. Use the same denominator.

$$\frac{2}{3} = \frac{10}{15}$$

$$+\frac{1}{5} = \frac{3}{15}$$

$$\frac{13}{15}$$

PRACTICE

Add. Simplify.

	a	b	c	d
1.	$\frac{1}{2} = \frac{9}{18}$ $+\frac{7}{9} = \frac{14}{18}$ $\frac{23}{18} = 1\frac{5}{18}$	$\frac{4}{5}$ $+\frac{2}{3}$	$\frac{1}{3}$ $+\frac{3}{10}$	$\frac{3}{7}$ $+\frac{1}{2}$
2.	$\frac{2}{3}$ $+\frac{3}{4}$	$\frac{5}{6}$ $+\frac{3}{5}$	$\frac{3}{7}$ $+\frac{1}{4}$	$\frac{1}{2}$ $+\frac{2}{8}$
3.	$\frac{2}{5}$ $+\frac{7}{9}$	$\frac{4}{5}$ $+\frac{7}{11}$	$\frac{2}{3}$ $+\frac{7}{8}$	$\frac{1}{7}$ $+\frac{3}{8}$

Add. Simplify.

	a	b	c
4.	$\frac{3}{8} + \frac{5}{6} = $ _____ $\frac{3}{8}$ $+\frac{5}{6}$	$\frac{5}{6} + \frac{3}{4} = $ _____	$\frac{2}{7} + \frac{2}{3} = $ _____

Adding Mixed Numbers, Whole Numbers, and Fractions

When adding mixed numbers, whole numbers, and fractions, first check for unlike denominators. Write mixed numbers and fractions as equivalent fractions with like denominators. Add the fractions. Then add the whole numbers and simplify.

Find: $3\frac{1}{6} + \frac{3}{4}$

Write the fractions with like denominators.	Add the fractions.	Add the whole numbers.
$3\frac{1}{6} = 3\frac{2}{12}$ $+ \ \frac{3}{4} = \frac{9}{12}$	$3\frac{1}{6} = 3\frac{2}{12}$ $+ \ \frac{3}{4} = \frac{9}{12}$ $\frac{11}{12}$	$3\frac{1}{6} = 3\frac{2}{12}$ $+ \ \frac{3}{4} = \frac{9}{12}$ $3\frac{11}{12}$

PRACTICE

Add. Simplify.

	a	b	c	d
1.	$5\frac{1}{3} = 5\frac{2}{6}$ $+ \ \frac{1}{6} = \frac{1}{6}$ $5\frac{3}{6} = 5\frac{1}{2}$	$\frac{2}{5}$ $+ 2\frac{1}{10}$	$3\frac{1}{7}$ $+ \ \frac{3}{14}$	$\frac{1}{6}$ $+ 9\frac{1}{2}$
2.	$11\frac{4}{9} = 11\frac{4}{9}$ $+ \ 4\frac{1}{3} = 4\frac{3}{9}$ $15\frac{7}{9}$	$2\frac{1}{4}$ $+ 3\frac{1}{8}$	$1\frac{1}{2}$ $+ 4\frac{3}{8}$	$6\frac{4}{15}$ $+ 3\frac{1}{3}$
3.	$6\frac{1}{8}$ $+ 2\frac{4}{7}$	$1\frac{1}{4}$ $+ 5\frac{1}{3}$	$3\frac{1}{2}$ $+ 6\frac{2}{5}$	$12\frac{1}{5}$ $+ \ 7\frac{3}{4}$
4.	5 $\frac{1}{10}$ $+ 2\frac{1}{2}$	$\frac{1}{4}$ $3\frac{2}{3}$ $+ 1$	$4\frac{1}{6}$ 6 $+ \ \frac{3}{8}$	7 $\frac{1}{2}$ $+ 8\frac{2}{5}$

PROBLEM-SOLVING STRATEGY

Use Guess and Check

A good way to solve some problems is to guess the answer. Then check it and make another guess if necessary. Your next guess will be better because you will learn from the first guess. Guess and check until you find the correct answer.

EXAMPLE 1
Read the problem.

Noriko made two telephone calls. The calls lasted a total of 15 minutes. The second call lasted 2 minutes longer than the first. How long was each call?

Decide what you already know.

One call was 2 minutes longer than the other.

Guess and check.

First Guess	*Second Guess*	*Third Guess*
1st call = 7 min.	1st call = 6 min.	1st call = $6\frac{1}{2}$ min.
2nd call = 9 min. (7 + 2)	2nd call = 8 min. (6 + 2)	2nd call = $8\frac{1}{2}$ min. ($6\frac{1}{2}$ + 2)
$\overline{16}$	$\overline{14}$	$\overline{15}$
Check: 16 is too many.	Check: 14 is too few.	Check: 15 is correct.

The calls lasted $6\frac{1}{2}$ and $8\frac{1}{2}$ minutes.

EXAMPLE 2
Read the problem.

The sum of the ages of Alex and his father is 60 years. Alex is 24 years younger than his father. How old is each?

Decide what you already know.

The father is 24 years older than Alex.

Guess and check.

First Guess	*Second Guess*
Alex - 20 years	Alex - 18 years
Father - 44 years (20 + 24)	Father - 42 years (18 + 24)
$\overline{64\text{ years}}$	$\overline{60\text{ years}}$
Check: 64 is too many.	Check: 60 is correct.

Alex is 18. His father is 42.

Solve. Use guess and check.

1. One day Pat Unger worked for a total of 8 hours. She worked 3 hours more in the afternoon than she worked in the morning. How long did she work in the afternoon?

Answer _____

2. The sum of the ages of Denise and Earl is 42 years. Earl is 8 years younger than Denise. How old is each?

Denise _____

Earl _____

3. Cheryl bought 2 bags of potatoes. The first bag weighed twice as much as the second. The bags weighed 20 pounds altogether. Find the weight of each bag.

1st bag _____

2nd bag _____

4. In the basketball game, Chet scored 37 points on 2-point and 3-point goals. He scored 6 more 2-pointers than 3-pointers. How many of each did he score?

2-pointers _____

3-pointers _____

5. What two numbers have a sum of 123 and a difference of 7?

Answer _____

6. What two numbers between 20 and 40 have a sum of 55 and a product of 736?

Answer _____

71

THE MEANING AND USE OF FRACTIONS

Adding Mixed Numbers, Whole Numbers, and Fractions with Large Sums

When adding mixed numbers, whole numbers, and fractions, your sum might contain an improper fraction. To regroup a sum that contains an improper fraction, first write the improper fraction as a mixed number. Then add and simplify.

Find: $8\frac{2}{3} + 2\frac{5}{6}$

Write the fractions with like denominators. Add.

$$8\frac{2}{3} = 8\frac{4}{6}$$
$$+2\frac{5}{6} = 2\frac{5}{6}$$
$$\overline{\qquad 10\frac{9}{6}}$$

The sum, $10\frac{9}{6}$, contains an improper fraction. To regroup, write the improper fraction as a mixed number.

$$\frac{9}{6} = 1\frac{1}{2}$$

Then add.

$$10\frac{9}{6} = 10 + 1\frac{1}{2} = 11\frac{1}{2}$$

PRACTICE

Add. Simplify.

	a	*b*	*c*

1.
$$3\frac{11}{12} = 3\frac{11}{12}$$
$$+\ \frac{1}{2} = \ \frac{6}{12}$$
$$\overline{3\frac{17}{12} = 3 + 1\frac{5}{12} = 4\frac{5}{12}}$$

    b: $\frac{5}{6}$ $+9\frac{3}{4}$

    c: $2\frac{9}{11}$ $+\ \frac{5}{8}$

2. $1\frac{3}{7}$ $+8\frac{7}{8}$ $6\frac{5}{6}$ $+4\frac{1}{2}$ $7\frac{8}{9}$ $+3\frac{2}{3}$

3. 5 $\frac{1}{2}$ $+\ \frac{2}{3}$ $\frac{5}{9}$ 8 $+\ \frac{1}{2}$ $\frac{3}{4}$ $\frac{2}{3}$ $+7$

4. $1\frac{1}{2}$ $3\frac{5}{8}$ $+\ \frac{3}{8}$ $9\frac{7}{10}$ $\frac{3}{10}$ $+4\frac{4}{5}$ $\frac{2}{3}$ $6\frac{5}{12}$ $+2\frac{1}{3}$

PROBLEM SOLVING

Applications

Solve.

1. Ann lives $\frac{7}{10}$ mile from town. Pam lives only $\frac{1}{10}$ mile from town. How much farther from town does Ann live than Pam?

Answer _____

2. Victor bought a watermelon that weighed $12\frac{3}{4}$ pounds. Luis bought one that weighed $2\frac{1}{4}$ pounds less. How much did Luis's watermelon weigh?

Answer _____

3. Martha had a board measuring $38\frac{5}{8}$ inches long. She cut off $19\frac{3}{8}$ inches so that the board would fit a shelf. How long was the board after being cut?

Answer _____

4. George bought a ham weighing $12\frac{7}{8}$ pounds. His brother cut off one slice. The slice weighed $1\frac{1}{8}$ pounds. How many pounds of ham were left?

Answer _____

5. One year, Milwaukee had a total snowfall of $88\frac{1}{2}$ inches. The normal snowfall is $52\frac{1}{4}$ inches. How much above normal was the snowfall?

Answer _____

6. The storekeeper sold $9\frac{1}{3}$ yards from a bolt of material containing $53\frac{2}{3}$ yards. How much material was left?

Answer _____

7. Last year David weighed $114\frac{1}{2}$ pounds. Now he weighs $119\frac{1}{2}$ pounds. How much weight has he gained?

Answer _____

8. For the first four months of the year, rain fell as follows: $2\frac{1}{2}$ inches, 3 inches, $1\frac{1}{2}$ inches, and $1\frac{1}{2}$ inches. What was the total rainfall for these four months?

Answer _____

Subtraction of Fractions with Different Denominators

To subtract fractions with different denominators, first rewrite the fractions as equivalent fractions with like denominators. Then subtract and simplify the answer.

Find: $\frac{5}{6} - \frac{3}{8}$

Write equivalent fractions with like denominators. Use the LCD.

$$\frac{5}{6} = \frac{5 \times 4}{6 \times 4} = \frac{20}{24}$$
$$-\frac{3}{8} = \frac{3 \times 3}{8 \times 3} = \frac{9}{24}$$

Subtract the numerators. Use the same denominator.

$$\frac{5}{6} = \frac{20}{24}$$
$$-\frac{3}{8} = \frac{9}{24}$$
$$\frac{11}{24}$$

GUIDED PRACTICE

Subtract. Simplify.

	a	*b*	*c*	*d*
1.	$\frac{1}{2} = \frac{5}{10}$ $-\frac{3}{10} = \frac{3}{10}$ $\frac{2}{10} = \frac{1}{5}$	$\frac{1}{3} = \frac{}{6}$ $-\frac{1}{6} = \frac{}{6}$	$\frac{5}{8} = \frac{}{16}$ $-\frac{5}{16} = \frac{}{16}$	$\frac{1}{4} = \frac{}{20}$ $-\frac{1}{5} = \frac{}{20}$
2.	$\frac{7}{10} = \frac{}{70}$ $-\frac{3}{7} = \frac{}{70}$	$\frac{1}{3} = \frac{}{12}$ $-\frac{1}{4} = \frac{}{12}$	$\frac{3}{10} = \frac{}{10}$ $-\frac{1}{5} = \frac{}{10}$	$\frac{1}{2} = \frac{}{12}$ $-\frac{1}{12} = \frac{}{12}$
3.	$\frac{3}{5} = \frac{}{35}$ $-\frac{2}{7} = \frac{}{35}$	$\frac{7}{8} = \frac{}{40}$ $-\frac{3}{10} = \frac{}{40}$	$\frac{5}{6} = \frac{}{12}$ $-\frac{3}{4} = \frac{}{12}$	$\frac{3}{4} = \frac{}{8}$ $-\frac{1}{8} = \frac{}{8}$
4.	$\frac{2}{3} = \frac{}{24}$ $-\frac{3}{8} = \frac{}{24}$	$\frac{5}{6} = \frac{}{6}$ $-\frac{1}{3} = \frac{}{6}$	$\frac{4}{5} = \frac{}{15}$ $-\frac{2}{3} = \frac{}{15}$	$\frac{4}{5} = \frac{}{10}$ $-\frac{7}{10} = \frac{}{10}$

PRACTICE

Subtract. Simplify.

	a	b	c	d
1.	$\frac{11}{12} = \frac{22}{24}$	$\frac{7}{9}$	$\frac{5}{6}$	$\frac{9}{11}$
	$-\frac{3}{8} = \frac{9}{24}$	$-\frac{1}{4}$	$-\frac{3}{8}$	$-\frac{3}{5}$
	$\frac{13}{24}$			

	a	b	c	d
2.	$\frac{1}{2}$	$\frac{8}{9}$	$\frac{4}{5}$	$\frac{14}{15}$
	$-\frac{1}{5}$	$-\frac{3}{8}$	$-\frac{3}{10}$	$-\frac{1}{3}$

	a	b	c	d
3.	$\frac{11}{12}$	$\frac{5}{6}$	$\frac{5}{7}$	$\frac{3}{4}$
	$-\frac{1}{3}$	$-\frac{5}{9}$	$-\frac{1}{2}$	$-\frac{3}{8}$

Subtract. Simplify.

	a	b	c
4.	$\frac{11}{12} - \frac{3}{8} = $ _____	$\frac{7}{8} - \frac{1}{4} = $ _____	$\frac{4}{5} - \frac{1}{2} = $ _____

$\frac{11}{12}$

$-\frac{3}{8}$

MIXED PRACTICE

Find each answer.

	a	b	c	d
1.	1 2 4 9	2 4,1 0 2	$5\frac{1}{3}$	$4\frac{3}{5}$
	6 3 4	1 9 0 8	$+8\frac{1}{12}$	$+6\frac{1}{10}$
	2 9 1 7	4 7 3 9		
	$+6 2 3 6$	$+\ \ \ \ 6 4 2$		

	a	b	c	d
2.	1 9	3 1 8	$5\,2\overline{)5\,7,7\,2\,0}$	$2\,6\overline{)2\,0\,7\,5}$
	$\times 3\,9$	$\times\ \ 2\,3$		

THE MEANING AND USE OF FRACTIONS

Subtraction of Fractions and Mixed Numbers from Whole Numbers

Sometimes you will need to subtract a fraction from a whole number. To subtract from a whole number, write the whole number as a mixed number with a like denominator. Then subtract the fractions. Subtract the whole numbers.

Find: $6 - 4\frac{1}{4}$

To subtract, you need two fractions with like denominators.	Write 6 as a mixed number with 4 as the denominator.	Subtract the fractions.	Subtract the whole numbers.
$\begin{array}{r} 6 \\ -4\frac{1}{4} \\ \hline \end{array}$	$6 = 5 + \frac{4}{4} = 5\frac{4}{4}$ Remember, $\frac{4}{4} = 1$	$\begin{array}{r} 6 = 5\frac{4}{4} \\ -4\frac{1}{4} = 4\frac{1}{4} \\ \hline \frac{3}{4} \end{array}$	$\begin{array}{r} 6 = 5\frac{4}{4} \\ -4\frac{1}{4} = 4\frac{1}{4} \\ \hline 1\frac{3}{4} \end{array}$

GUIDED PRACTICE

Write each whole number as a mixed number.

	a	b	c	d
1.	$12 = 11 + \frac{2}{2} = 11\frac{2}{2}$	$3 = 2 + \frac{}{9} =$	$21 = 20 + \frac{}{10} =$	$42 = 41 + \frac{}{24} =$
2.	$7 = 6 + \frac{}{5} =$	$14 = 13 + \frac{}{7} =$	$16 = 15 + \frac{}{11} =$	$35 = 34 + \frac{}{6} =$

Subtract.

	a	b	c	d
3.	$\begin{array}{r} 9 = 8\frac{5}{5} \\ -1\frac{2}{5} = 1\frac{2}{5} \\ \hline 7\frac{3}{5} \end{array}$	$\begin{array}{r} 7 = 6\frac{}{4} \\ -2\frac{3}{4} = 2\frac{3}{4} \\ \hline \end{array}$	$\begin{array}{r} 11 = 10\frac{}{6} \\ -8\frac{5}{6} = 8\frac{5}{6} \\ \hline \end{array}$	$\begin{array}{r} 4 = 3\frac{}{3} \\ -2\frac{1}{3} = 2\frac{1}{3} \\ \hline \end{array}$
4.	$\begin{array}{r} 10 = 9\frac{11}{11} \\ -2\frac{5}{11} = 2\frac{5}{11} \\ \hline 7\frac{6}{11} \end{array}$	$\begin{array}{r} 6 = \\ -4\frac{2}{3} = \\ \hline \end{array}$	$\begin{array}{r} 9 = \\ -1\frac{2}{7} = \\ \hline \end{array}$	$\begin{array}{r} 12 = \\ -6\frac{5}{6} = \\ \hline \end{array}$
5.	$\begin{array}{r} 8 = 7\frac{3}{3} \\ -1\frac{1}{3} = \frac{1}{3} \\ \hline 7\frac{2}{3} \end{array}$	$\begin{array}{r} 2 = \\ -\frac{3}{5} = \\ \hline \end{array}$	$\begin{array}{r} 5 = \\ -\frac{1}{2} = \\ \hline \end{array}$	$\begin{array}{r} 3 = \\ -\frac{1}{4} = \\ \hline \end{array}$

PRACTICE

Subtract.

	a	b	c	d
1.	10	2	8	25
	$-\ 4\frac{5}{6}$	$-\ \frac{2}{7}$	$-\ 1\frac{1}{6}$	$-\ 12\frac{5}{7}$

	a	b	c	d
2.	6	5	21	3
	$-\ 2\frac{4}{9}$	$-\ 1\frac{1}{4}$	$-\ 3\frac{1}{9}$	$-\ \frac{9}{16}$

	a	b	c	d
3.	19	12	5	14
	$-\ 13\frac{5}{6}$	$-\ \frac{1}{3}$	$-\ 3\frac{3}{8}$	$-\ 11\frac{2}{5}$

	a	b	c	d
4.	3	10	8	18
	$-\ 2\frac{7}{10}$	$-\ 7\frac{1}{5}$	$-\ \frac{3}{5}$	$-\ 13\frac{5}{8}$

Line up the digits. Then subtract.

a	b	c
5. $10 - \frac{3}{4} = $ _____	$6 - \frac{1}{5} = $ _____	$11 - 4\frac{7}{12} = $ _____

$$\begin{array}{r} 10 \\ -\ \frac{3}{4} \\ \hline \end{array}$$

a	b	c
6. $12 - 9\frac{5}{9} = $ _____	$7 - 5\frac{1}{2} = $ _____	$9 - \frac{3}{10} = $ _____

⇒ **MIXED PRACTICE**

Find each answer.

	a	b	c	d
1.	$45\overline{)4700}$	$\begin{array}{r} 2407 \\ \times\ \ 908 \\ \hline \end{array}$	$38\overline{)9728}$	$\begin{array}{r} 983,451 \\ 80,169 \\ +\ 619,242 \\ \hline \end{array}$

THE MEANING AND USE OF FRACTIONS

Subtraction of Mixed Numbers with Regrouping

When subtracting mixed numbers, it may be necessary to regroup first. To regroup a mixed number for subtraction, write the whole number part as a mixed number. Add the mixed number and the fraction. Then subtract and simplify.

Find: $5\frac{1}{8} - 3\frac{3}{4}$

Write the fractions with like denominators. Compare the numerators.	$\frac{6}{8}$ is bigger than $\frac{1}{8}$. You can't subtract the fractions. To regroup, write 5 as a mixed number.	Add the mixed number and the fraction.	Now you can subtract.
$5\frac{1}{8} = 5\frac{1}{8}$ $-3\frac{3}{4} = 3\frac{6}{8}$	$5 = 4\frac{8}{8}$	$5\frac{1}{8} = 4\frac{8}{8} + \frac{1}{8} = 4\frac{9}{8}$ Remember, $\frac{9}{8}$ is an improper fraction.	$5\frac{1}{8} = 4\frac{9}{8}$ $-3\frac{3}{4} = 3\frac{6}{8}$ $1\frac{3}{8}$

GUIDED PRACTICE

Regroup each mixed number.

a	b	c
1. $9\frac{1}{4} = 8\frac{4}{4} + \frac{1}{4} = 8\frac{5}{4}$	$3\frac{2}{5} =$	$12\frac{1}{8} =$
2. $5\frac{1}{3} =$	$6\frac{3}{7} =$	$7\frac{5}{6} =$

Subtract. Simplify.

a	b	c
3. $7\frac{1}{4} = 6\frac{5}{4}$ $-3\frac{3}{4} = 3\frac{3}{4}$ $3\frac{2}{4} = 3\frac{1}{2}$	$3\frac{1}{3} = 2\frac{\ }{3}$ $-\ \frac{2}{3} = \ \frac{2}{3}$	$5\frac{3}{10} = 4\frac{\ }{10}$ $-2\frac{7}{10} = 2\frac{7}{10}$
4. $6\frac{1}{9} = 6\frac{1}{9} = 5\frac{10}{9}$ $-\ \frac{2}{3} = \ \frac{6}{9} = \ \frac{6}{9}$ $5\frac{4}{9}$	$7\frac{1}{8} = 7\frac{1}{8} =$ $-5\frac{1}{2} = 5\frac{4}{8} =$	$10\frac{1}{6} = 10\frac{2}{12} =$ $-\ 4\frac{5}{12} = \ 4\frac{5}{12} =$
5. $8\frac{1}{12} = 8\frac{3}{36} = 7\frac{39}{36}$ $-4\frac{8}{9} = 4\frac{32}{36} = 4\frac{32}{36}$ $3\frac{7}{36}$	$15\frac{1}{6} = 15\frac{2}{12} =$ $-12\frac{3}{4} = 12\frac{9}{12} =$	$5\frac{1}{6} = 5\frac{4}{24} =$ $-\ \frac{7}{8} = \ \frac{21}{24} =$

PRACTICE

Subtract. Simplify.

	a	b	c
1.	$6\frac{2}{5}$	$5\frac{1}{3}$	$7\frac{1}{10}$
	$-4\frac{1}{2}$	$-3\frac{2}{3}$	$-\ \ \frac{1}{5}$

	a	b	c
2.	$12\frac{1}{8}$	$9\frac{1}{5}$	$11\frac{1}{4}$
	$-\ \ 9\frac{5}{8}$	$-6\frac{5}{8}$	$-\ 5\frac{3}{7}$

	a	b	c
3.	$8\frac{3}{5}$	$2\frac{1}{3}$	$15\frac{1}{8}$
	$-4\frac{9}{10}$	$-\ \ \frac{2}{3}$	$-\ \ \frac{3}{4}$

Line up the digits. Subtract. Simplify.

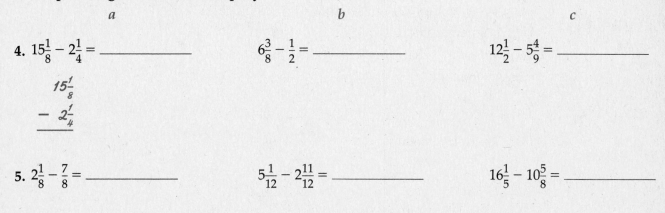

a b c

4. $15\frac{1}{8} - 2\frac{1}{4} =$ _____ $6\frac{3}{8} - \frac{1}{2} =$ _____ $12\frac{1}{2} - 5\frac{4}{9} =$ _____

$$15\tfrac{1}{8}$$
$$-\ 2\tfrac{1}{4}$$

5. $2\frac{1}{8} - \frac{7}{8} =$ _____ $5\frac{1}{12} - 2\frac{11}{12} =$ _____ $16\frac{1}{5} - 10\frac{5}{8} =$ _____

> **MIXED PRACTICE**
>
> **Estimate each sum or difference.**

	a	b	c	d
1.	$581 \rightarrow$	$5{,}176$	$21{,}875$	6508
	$+487 \rightarrow$	$+17{,}464$	$-\ \ \ 799$	$-\ \ 502$

PROBLEM-SOLVING STRATEGY

Make a Drawing

When you read a problem, you might know at once how to solve it. Sometimes, however, the solution is not obvious. When that happens, a drawing may help you solve the problem. You should always read the problem carefully and look for facts. Next, make a drawing to help you keep track of the facts. Then, solve the problem.

Read the problem.

Chuck is $\frac{1}{2}$ foot taller than Amy. Brenda is $\frac{2}{3}$ foot shorter than Chuck. Brenda is $5\frac{1}{3}$ feet tall. How tall are Amy and Chuck?

List the facts.

Fact 1. Brenda is $5\frac{1}{3}$ feet tall.

Fact 2. Brenda is $\frac{2}{3}$ foot shorter than Chuck.

Fact 3. Chuck is $\frac{1}{2}$ foot taller than Amy.

Make a drawing.

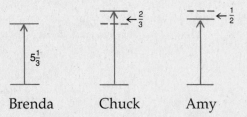

Brenda   Chuck   Amy

Solve the problem.

From the drawing you can see that Chuck's height is Brenda's height plus $\frac{2}{3}$ foot.

You can also see that Amy's height is Chuck's height minus $\frac{1}{2}$ foot.

Chuck's height

$5\frac{1}{3}$ Brenda's height

$+\ \dfrac{2}{3}$

$\overline{\ \ 5\frac{3}{3}}$ = 6 feet

Amy's height

$6 = 5\frac{2}{2}$ Chuck's height

$-\ \dfrac{1}{2} = \dfrac{1}{2}$

$\overline{\qquad 5\frac{1}{2}}$ feet

Chuck is 6 feet tall. Amy is $5\frac{1}{2}$ feet tall.

For each of the following problems, make a drawing or diagram. Write in all the facts. Then solve the problem.

1. Ann is taller than Bob but shorter than Carlos. David is taller than Bob but shorter than Ann. Elena is taller than Ann but shorter than Carlos. List the names from shortest to tallest.

Answer _____

2. Ron told a secret to 4 people. Each of those people told the secret to 2 more people. Each of them told 1 other person. How many people now know the secret?

Answer _____

3. A snail crawls $5\frac{3}{4}$ inches north, $2\frac{1}{2}$ inches east, $6\frac{1}{2}$ inches south, $4\frac{1}{4}$ inches west, and $\frac{3}{4}$ inch north. How far is it from its starting point?

Answer _____

4. The top of a mountain is 1470 feet below a cloud. The mountain is 12,805 feet above sea level. An eagle flies 7516 feet above sea level. How far below the cloud is the eagle?

Answer _____

5. Lou rode his bike 15 miles north from his house and got a flat tire. He walked $\frac{3}{4}$ mile west to a phone booth to call Pat who lives 15 miles south of the phone booth. How far from Lou's house is Pat's house?

Answer _____

6. Train A leaves the station at 9:00 and travels at 60 miles an hour. Train B leaves the station at 9:15 and travels at 75 miles an hour. Which train will be farther from the station at 10:00?

Answer _____

PROBLEM SOLVING

Applications

Solve.

1. One bottle of cologne contains $\frac{1}{2}$ ounce. A smaller bottle contains $\frac{5}{16}$ ounce. How much more cologne does the larger bottle contain?

Answer _____

2. Ginger needs a sheet of paper that is $7\frac{3}{4}$ inches wide. She has a piece that is $8\frac{7}{8}$ inches wide. How much does she need to cut off?

Answer _____

3. A share of Maxiflex stock costs $8\frac{7}{8}$ dollars. A share of Bentley stock sells for $8\frac{3}{8}$ dollars. Which stock costs less? How much less?

Answer _____

Answer _____

4. Frank lives 8 miles from town. Hector lives $\frac{5}{8}$ mile closer to town. How far from town does Hector live?

Answer _____

5. On Monday, $\frac{1}{2}$ inch of rain fell. On Tuesday, $\frac{3}{4}$ inch fell. On Wednesday, it rained a total of $\frac{1}{8}$ inch. How much rain fell on the three days?

Answer _____

6. Diana bought a turkey that weighs $22\frac{1}{4}$ pounds. Her brother Joe bought one that weighs $17\frac{1}{2}$ pounds. Diana's turkey weighs how much more than Joe's?

Answer _____

7. Louisa has a ring that is $\frac{3}{8}$ inch wide. Jeff has a ring that is $\frac{3}{4}$ inch wide. How much wider is Jeff's ring than Louisa's ring?

Answer _____

8. The Smiths, the Lees, and the Greens live outside of town. The Smiths live 5 miles from town. The Lees live $\frac{3}{4}$ mile farther. The Greens live $\frac{1}{2}$ mile beyond the Lees. How far from town do the Greens live?

Answer _____

THE MEANING AND USE OF FRACTIONS

Unit 3 Review

Write as improper fractions.

a	*b*	*c*	*d*
1. $4\frac{1}{3} =$	$5\frac{2}{3} =$	$6\frac{2}{5} =$	$8\frac{5}{8} =$

Write as mixed numbers or whole numbers. Simplify.

a	*b*	*c*	*d*
2. $\frac{17}{3} =$	$\frac{19}{4} =$	$\frac{31}{5} =$	$\frac{47}{9} =$
3. $\frac{45}{15} =$	$\frac{40}{18} =$	$\frac{135}{45} =$	$\frac{52}{10} =$

Write as equivalent fractions.

a	*b*	*c*	*d*
4. $\frac{2}{3} = \frac{}{9}$	$\frac{3}{4} = \frac{}{16}$	$\frac{4}{5} = \frac{}{35}$	$\frac{7}{10} = \frac{}{20}$

Add or subtract. Simplify.

a	*b*	*c*	*d*
5. $\frac{2}{3}$ $+\frac{1}{3}$	$\frac{1}{2}$ $+\frac{1}{2}$	$\frac{4}{5}$ $+\frac{3}{5}$	$\frac{5}{6}$ $+\frac{5}{6}$
6. $\frac{9}{10}$ $+\frac{2}{3}$	$5\frac{3}{5}$ $+3\frac{3}{4}$	$\frac{3}{5}$ $-\frac{1}{5}$	$1\,2$ $-\;2\frac{2}{5}$
7. $\frac{7}{8}$ $+\frac{1}{2}$	$8\frac{1}{3}$ $-2\frac{2}{5}$	$\frac{15}{16}$ $-\frac{3}{16}$	$\frac{2}{5}$ $+6\frac{1}{4}$

Solve.

8. Mary needs $2\frac{2}{3}$ yards of cloth for a dress. She has $3\frac{3}{4}$ yards. How much cloth will she have left?

Answer _____

9. Leroy ran $2\frac{1}{4}$ miles on one day. He ran 3 miles the next day and $3\frac{2}{5}$ miles the third day. How many miles did he run?

Answer _____

Multiplication of Fractions

To multiply fractions, multiply the numerators and multiply the denominators. Simplify the answer.

Find: $\frac{1}{7} \times \frac{4}{5}$

Multiply the numerators.

$$\frac{1}{7} \times \frac{4}{5} = \frac{1 \times 4}{} = \frac{4}{}$$

Multiply the denominators.

$$\frac{1}{7} \times \frac{4}{5} = \frac{1 \times 4}{7 \times 5} = \frac{4}{35}$$

Find: $\frac{2}{3} \times \frac{3}{8}$

Multiply the numerators.

$$\frac{2}{3} \times \frac{3}{8} = \frac{2 \times 3}{} = \frac{6}{}$$

Multiply the denominators.
Simplify.

$$\frac{2}{3} \times \frac{3}{8} = \frac{2 \times 3}{3 \times 8} = \frac{6}{24} = \frac{1}{4}$$

PRACTICE

Multiply. Simplify.

  a   *b*

1. $\frac{2}{3} \times \frac{4}{5} = \frac{2 \times 4}{3 \times 5} = \frac{8}{15}$ $\frac{4}{5} \times \frac{2}{7} =$

2. $\frac{1}{2} \times \frac{1}{3} =$ $\frac{2}{3} \times \frac{1}{5} =$

3. $\frac{3}{4} \times \frac{7}{8} =$ $\frac{4}{9} \times \frac{4}{5} =$

4. $\frac{3}{5} \times \frac{2}{3} =$ $\frac{3}{10} \times \frac{5}{8} =$

5. $\frac{2}{3} \times \frac{1}{2} =$ $\frac{3}{5} \times \frac{2}{9} =$

6. $\frac{3}{7} \times \frac{2}{3} =$ $\frac{5}{6} \times \frac{2}{5} =$

7. $\frac{2}{3} \times \frac{1}{4} =$ $\frac{5}{8} \times \frac{2}{5} =$

MULTIPLICATION AND DIVISION OF FRACTIONS

Multiplication of Fractions Using Cancellation

Instead of simplifying fractions after they have been multiplied, it may be possible to use *cancellation* before multiplying. To cancel, find a common factor of a numerator and a denominator. Divide the numerator and the denominator by the common factor. Then multiply, using the new numerator and denominator.

Find: $\frac{3}{10} \times \frac{1}{9}$

Find the common factor.	Cancel.	Multiply the new numerators and denominators.
$\frac{3}{10} \times \frac{1}{9}$	$\overset{1}{\frac{\cancel{3}}{10}} \times \frac{1}{\underset{3}{\cancel{9}}}$	$\frac{1 \times 1}{10 \times 3} = \frac{1}{30}$
The common factor of 3 and 9 is 3.	Divide both the 3 and the 9 by 3.	

PRACTICE

Multiply using cancellation.

a b

1. $\frac{1}{\underset{5}{\cancel{10}}} \times \frac{\overset{2}{\cancel{4}}}{7} = \frac{1 \times 2}{5 \times 7} = \frac{2}{35}$ $\frac{3}{35} \times \frac{5}{8} =$

2. $\frac{3}{4} \times \frac{4}{5} =$ $\frac{2}{3} \times \frac{6}{7} =$

3. $\frac{5}{11} \times \frac{3}{10} =$ $\frac{1}{4} \times \frac{2}{5} =$

4. $\frac{1}{3} \times \frac{3}{7} =$ $\frac{3}{5} \times \frac{10}{13} =$

5. $\frac{\overset{1}{\cancel{7}}}{\underset{2}{\cancel{8}}} \times \frac{\overset{1}{\cancel{4}}}{\underset{3}{\cancel{21}}} = \frac{1 \times 1}{2 \times 3} = \frac{1}{6}$ $\frac{3}{4} \times \frac{8}{9} =$

6. $\frac{5}{9} \times \frac{3}{10} =$ $\frac{7}{8} \times \frac{2}{7} =$

7. $\frac{5}{12} \times \frac{4}{15} =$ $\frac{3}{7} \times \frac{7}{9} =$

8. $\frac{3}{8} \times \frac{4}{9} =$ $\frac{5}{8} \times \frac{2}{5} =$

MULTIPLICATION AND DIVISION OF FRACTIONS

Multiplication of Whole Numbers by Fractions

To multiply a whole number by a fraction, first write the whole number as an improper fraction. Use cancellation if possible. Multiply the new numerators and denominators. Simplify the answer.

Find: $\frac{4}{9} \times 3$

Write the whole number as an improper fraction.	Cancel.	Multiply the new numerators and denominators.	Simplify.
$\frac{4}{9} \times 3 = \frac{4}{9} \times \frac{3}{1}$	$\frac{4}{\underset{3}{\cancel{9}}} \times \frac{\overset{1}{\cancel{3}}}{1}$	$\frac{4 \times 1}{3 \times 1} = \frac{4}{3}$	$\frac{4}{3} = 1\frac{1}{3}$

GUIDED PRACTICE

Write each whole number as an improper fraction.

	a	b	c	d
1.	$1 = \dfrac{1}{1}$	$27 = \underline{\hphantom{xxx}}$	$19 = \underline{\hphantom{xxx}}$	$52 = \underline{\hphantom{xxx}}$
2.	$7 = \underline{\hphantom{xxx}}$	$36 = \underline{\hphantom{xxx}}$	$125 = \underline{\hphantom{xxx}}$	$11 = \underline{\hphantom{xxx}}$

Multiply using cancellation. Simplify.

      a   b

3. $\frac{5}{6} \times 18 = \frac{5}{\underset{1}{\cancel{6}}} \times \frac{\overset{3}{\cancel{18}}}{1} = \frac{5 \times 3}{1 \times 1} = \frac{15}{1} = 15$ $\frac{2}{5} \times 20 = \frac{2}{5} \times \frac{20}{1} =$

4. $\frac{2}{3} \times 15 = \frac{2}{3} \times \frac{15}{1} =$ $\frac{2}{3} \times 4 = \frac{2}{3} \times \frac{4}{1} =$

5. $\frac{2}{5} \times 125 = \frac{2}{5} \times \frac{125}{1} =$ $\frac{3}{10} \times 50 = \frac{3}{10} \times \frac{50}{1} =$

6. $12 \times \frac{3}{4} = \frac{\overset{3}{\cancel{12}}}{1} \times \frac{3}{\underset{1}{\cancel{4}}} = \frac{3 \times 3}{1 \times 1} = \frac{9}{1} = 9$ $18 \times \frac{3}{4} = \frac{18}{1} \times \frac{3}{4} =$

7. $25 \times \frac{3}{15} = \frac{25}{1} \times \frac{3}{15} =$ $21 \times \frac{3}{7} = \frac{21}{1} \times \frac{3}{7} =$

8. $9 \times \frac{3}{4} = \frac{9}{1} \times \frac{3}{4} =$ $18 \times \frac{2}{9} = \frac{18}{1} \times \frac{2}{9} =$

PRACTICE

Multiply using cancellation. Simplify.

	a	b
1.	$6 \times \frac{4}{5} =$	$\frac{1}{3} \times 21 =$
2.	$24 \times \frac{5}{8} =$	$7 \times \frac{1}{2} =$
3.	$75 \times \frac{2}{5} =$	$10 \times \frac{3}{5} =$
4.	$\frac{2}{3} \times 4 =$	$24 \times \frac{5}{8} =$
5.	$33 \times \frac{1}{3} =$	$15 \times \frac{1}{5} =$
6.	$40 \times \frac{5}{8} =$	$\frac{5}{6} \times 30 =$
7.	$2 \times \frac{2}{7} =$	$32 \times \frac{3}{8} =$
8.	$\frac{5}{12} \times 36 =$	$\frac{1}{5} \times 45 =$

→ MIXED PRACTICE

Find each answer.

	a	b	c	d
1.	3792 +1864	705 −239	15,317 + 892	7363 −1497
2.	437 × 29	36)1476	247 ×129	2316 4219 1657 + 702

MULTIPLICATION AND DIVISION OF FRACTIONS

Multiplication of Mixed Numbers by Whole Numbers

To multiply a mixed number by a whole number, write the mixed number and the whole number as improper fractions. Use cancellation if possible. Multiply the new numerators and denominators. Simplify the answer.

Find: $1\frac{3}{5} \times 15$

Write the whole number and the mixed number as improper fractions.	Cancel.	Multiply the new numerators and denominators. Simplify.
$1\frac{3}{5} \times 15 = \frac{8}{5} \times \frac{15}{1}$	$\frac{8}{\overset{}{\underset{1}{5}}} \times \frac{\overset{3}{15}}{1}$	$\frac{8 \times 3}{1 \times 1} = \frac{24}{1} = 24$

PRACTICE

Multiply. Use cancellation if possible. Simplify.

a

1. $1\frac{1}{2} \times 4 = \frac{3}{\underset{1}{\cancel{2}}} \times \frac{\overset{2}{\cancel{4}}}{1} = \frac{3 \times 2}{1 \times 1} = \frac{6}{1} = 6$

2. $2\frac{1}{3} \times 9 =$

3. $3\frac{14}{15} \times 2 =$

4. $12\frac{1}{10} \times 20 =$

5. $1\frac{2}{3} \times 6 =$

6. $4\frac{9}{10} \times 1 =$

7. $1\frac{3}{8} \times 14 =$

8. $2\frac{3}{4} \times 10 =$

b

$4\frac{3}{4} \times 3 =$

$1\frac{4}{5} \times 5 =$

$9\frac{1}{10} \times 11 =$

$4\frac{5}{6} \times 9 =$

$4\frac{5}{21} \times 6 =$

$2\frac{3}{10} \times 7 =$

$5\frac{1}{6} \times 2 =$

$1\frac{5}{6} \times 3 =$

MULTIPLICATION AND DIVISION OF FRACTIONS
Multiplication of Mixed Numbers by Fractions

To multiply a mixed number by a fraction, write the mixed number as an improper fraction. Use cancellation if possible. Multiply the new numerators and denominators. Simplify the answer.

Find: $\frac{2}{3} \times 5\frac{1}{4}$

Write the mixed number as an improper fraction.	Cancel.	Multiply the new numerators and denominators. Simplify.
$\frac{2}{3} \times 5\frac{1}{4} = \frac{2}{3} \times \frac{21}{4}$	$\frac{\cancel{2}^1}{\cancel{3}_1} \times \frac{\cancel{21}^7}{\cancel{4}_2}$	$\frac{1 \times 7}{1 \times 2} = \frac{7}{2} = 3\frac{1}{2}$

PRACTICE

Multiply. Use cancellation if possible. Simplify.

a

1. $\frac{1}{2} \times 3\frac{1}{2} = \frac{1}{2} \times \frac{7}{2} = \frac{1 \times 7}{2 \times 2} = \frac{7}{4} = 1\frac{3}{4}$

2. $\frac{2}{3} \times 5\frac{1}{4} =$

3. $\frac{1}{2} \times 1\frac{3}{5} =$

4. $\frac{12}{23} \times 5\frac{3}{4} =$

5. $3\frac{1}{2} \times \frac{3}{8} =$

6. $7\frac{2}{3} \times \frac{1}{2} =$

7. $2\frac{1}{2} \times \frac{1}{3} =$

8. $2\frac{1}{3} \times \frac{6}{7} =$

b

$\frac{3}{5} \times 3\frac{1}{3} =$

$\frac{2}{9} \times 4\frac{1}{2} =$

$\frac{1}{2} \times 3\frac{1}{2} =$

$\frac{9}{16} \times 2\frac{2}{3} =$

$1\frac{7}{8} \times \frac{4}{15} =$

$4\frac{2}{3} \times \frac{3}{7} =$

$9\frac{1}{2} \times \frac{1}{8} =$

$3\frac{3}{4} \times \frac{7}{12} =$

Multiplication of Mixed Numbers by Mixed Numbers

To multiply a mixed number by a mixed number, write both mixed numbers as improper fractions. Use cancellation if possible. Multiply the new numerators and denominators. Simplify the answer.

Find: $1\frac{2}{3} \times 4\frac{1}{2}$

Write the mixed numbers as improper fractions.	Cancel.	Multiply the new numerators and denominators. Simplify.
$1\frac{2}{3} \times 4\frac{1}{2} = \frac{5}{3} \times \frac{9}{2}$	$\dfrac{5}{\underset{1}{3}} \times \dfrac{\overset{3}{9}}{2}$	$\dfrac{5 \times 3}{1 \times 2} = \dfrac{15}{2} = 7\frac{1}{2}$

PRACTICE

Multiply. Use cancellation if possible. Simplify.

a

1. $2\frac{3}{8} \times 2\frac{1}{3} = \frac{19}{8} \times \frac{7}{3} = \frac{19 \times 7}{8 \times 3} = \frac{133}{24} = 5\frac{13}{24}$

2. $2\frac{2}{5} \times 4\frac{1}{2} =$

3. $1\frac{1}{14} \times 1\frac{3}{4} =$

4. $3\frac{4}{7} \times 2\frac{4}{5} =$

5. $8\frac{3}{9} \times 1\frac{2}{25} =$

6. $2\frac{1}{3} \times 5\frac{1}{5} =$

7. $1\frac{4}{5} \times 1\frac{2}{9} =$

8. $2\frac{2}{7} \times 1\frac{2}{5} =$

b

1. $1\frac{1}{3} \times 2\frac{1}{2} =$

2. $2\frac{1}{5} \times 2\frac{1}{4} =$

3. $3\frac{3}{4} \times 2\frac{2}{3} =$

4. $2\frac{3}{4} \times 5\frac{1}{4} =$

5. $6\frac{1}{2} \times 1\frac{1}{3} =$

6. $2\frac{1}{12} \times 3\frac{6}{25} =$

7. $2\frac{5}{6} \times 3\frac{3}{4} =$

8. $4\frac{3}{5} \times 2\frac{1}{3} =$

PROBLEM SOLVING

Applications

Solve.

1. Mr. Yanaga's car has a gas tank that holds 16 gallons. He used $\frac{3}{4}$ of a tank of gas. How many gallons of gas did he use?

   Answer _____

2. Margie had a bag of nails which weighed three fourths of a pound. She gave one half of the nails to Leona. How much did the bag weigh then?

   Answer _____

3. Terry lives $\frac{5}{8}$ mile from town. Patricia lives halfway between Terry and town. How far from town does Patricia live?

   Answer _____

4. Anna bought $\frac{3}{4}$ foot of heater hose at 80 cents per foot. How much did she pay?

   Answer _____

5. Kevin made five shirts. Each shirt required $\frac{2}{3}$ yard of cloth. How many yards did he need in all?

   Answer _____

6. Wilbur lives $\frac{9}{16}$ mile from his office. Leona lives only one third as far from her office. How far is it to Leona's office?

   Answer _____

7. Maria bought $25\frac{1}{2}$ feet of tubing at 40 cents a foot. How much did she pay for all of it?

   Answer _____

8. Lin wished to make T-shirts for each of her five children. Each shirt required $\frac{3}{4}$ yard of cloth. How many yards did she need in all?

   Answer _____

PROBLEM-SOLVING STRATEGY

Make a Table

You should always look for facts as you read a problem. When many facts are given in a problem, it can be helpful to organize the facts in a table. Then, you can use the table to solve the problem.

Read the problem.

Maloney's Pet Shop had 10 dogs and 26 cats. Each week they sold 1 cat and bought 3 dogs. In which week did Maloney's Pet Shop have the same number of cats and dogs?

List the facts.

Fact 1. Maloney's Pet Shop had 10 dogs and 26 cats.
Fact 2. Each week they sold 1 cat and bought 3 dogs.

Make a table.

	Week 1	Week 2	Week 3	Week 4	Week 5
Dogs	10	13	16	19	22
Cats	26	25	24	23	22

Solve the problem.

Look at the table.
After 5 weeks, Maloney's Pet Shop had 22 dogs and 22 cats.

Solve by making a table.

1. Nora was offered two jobs. The first job paid $20,000 a year with $750 annual increases. The second job paid $24,000 a year with $250 annual increases. In how many years will the annual salaries be the same?

2. Kevin started running the first week in January. He ran 4 miles 3 days a week. Greg started running the second week of January. He ran 3 miles 5 times a week. By the end of what week had they run the same distance?

Answer _____

Answer _____

Solve by making a table.

3. In 1975, the price of a Frosty refrigerator was $480. A Coldwell refrigerator sold for $545. The price of a Frosty went up $38 annually, while the Coldwell went up $25 annually. In what year did the two refrigerators sell for the same price?

Answer _____

4. Gretchen is a quality control engineer for a lawn mower factory. She checks every 12th mower for ease-of-starting and every 21st mower for ease-of-handling. How often does she check a mower for both features?

Answer _____

5. Vernon weighed 210 pounds and Henry weighed 155 pounds. Each month Vernon lost 8 pounds and Henry gained 3 pounds. How much did they weigh when their weights were equal?

Answer _____

6. Lawn fertilizer was sold in 3-pound and 7-pound bags. Mrs. Ikeda bought 35 pounds of fertilizer. How many bags of each size might she have bought?

Answer _____

MULTIPLICATION AND DIVISION OF FRACTIONS

Division of Fractions by Fractions

To divide a fraction by a fraction, multiply by the reciprocal of the second fraction. To find the reciprocal, invert the second fraction. For example, the reciprocal of $\frac{3}{4}$ is $\frac{4}{3}$. Simplify your answer, if needed.

Remember, only the second fraction is inverted.

Find: $\frac{2}{3} \div \frac{5}{12}$

Multiply by the reciprocal of the second fraction.	Cancel.	Multiply the new numerators and denominators. Simplify.
$\frac{2}{3} \div \frac{5}{12} = \frac{2}{3} \times \frac{12}{5}$	$\frac{2}{\underset{1}{3}} \times \frac{\overset{4}{12}}{5}$	$\frac{2 \times 4}{1 \times 5} = \frac{8}{5} = 1\frac{3}{5}$

GUIDED PRACTICE

Write the reciprocal.

	a	b	c	d	e
1.	$\frac{2}{9}$ ————	$\frac{3}{4}$ ————	$\frac{5}{6}$ ————	$\frac{7}{10}$ ————	$\frac{1}{8}$ ————
2.	$\frac{8}{9}$ ————	$\frac{1}{7}$ ————	$\frac{1}{2}$ ————	$\frac{2}{5}$ ————	$\frac{5}{11}$ ————

Divide. Simplify.

a

3. $\frac{4}{9} \div \frac{8}{9} = \frac{\overset{1}{4}}{\underset{1}{9}} \times \frac{\overset{1}{9}}{\underset{2}{8}} = \frac{1 \times 1}{1 \times 2} = \frac{1}{2}$

b

$\frac{5}{6} \div \frac{3}{4} = \frac{5}{6} \times \frac{4}{3} =$

4. $\frac{5}{8} \div \frac{1}{2} = \frac{5}{8} \times \frac{2}{1} =$

$\frac{7}{12} \div \frac{7}{8} = \frac{7}{12} \times \frac{8}{7} =$

5. $\frac{2}{8} \div \frac{3}{8} = \frac{2}{8} \times \frac{8}{3} =$

$\frac{15}{16} \div \frac{2}{3} = \frac{15}{16} \times \frac{3}{2} =$

6. $\frac{1}{3} \div \frac{1}{16} = \frac{1}{3} \times \frac{16}{1} =$

$\frac{4}{5} \div \frac{8}{15} = \frac{4}{5} \times \frac{15}{8} =$

7. $\frac{2}{3} \div \frac{3}{8} = \frac{2}{3} \times \frac{8}{3} =$

$\frac{8}{11} \div \frac{1}{2} = \frac{8}{11} \times \frac{2}{1} =$

8. $\frac{3}{4} \div \frac{5}{8} = \frac{3}{4} \times \frac{8}{5} =$

$\frac{9}{16} \div \frac{3}{8} = \frac{9}{16} \times \frac{8}{3} =$

Divide. Simplify.

        a   *b*

1. $\frac{1}{8} \div \frac{1}{3} =$ $\frac{5}{12} \div \frac{3}{4} =$

2. $\frac{9}{16} \div \frac{3}{8} =$ $\frac{1}{3} \div \frac{1}{7} =$

3. $\frac{7}{8} \div \frac{5}{12} =$ $\frac{4}{6} \div \frac{1}{10} =$

4. $\frac{1}{3} \div \frac{1}{3} =$ $\frac{10}{64} \div \frac{1}{4} =$

5. $\frac{2}{9} \div \frac{3}{4} =$ $\frac{3}{7} \div \frac{3}{14} =$

6. $\frac{3}{5} \div \frac{7}{8} =$ $\frac{4}{9} \div \frac{2}{3} =$

7. $\frac{5}{6} \div \frac{5}{8} =$ $\frac{1}{8} \div \frac{3}{16} =$

8. $\frac{5}{16} \div \frac{5}{32} =$ $\frac{1}{3} \div \frac{1}{5} =$

➡ MIXED PRACTICE

Find each answer.

     a   *b*   *c*   *d*

1.

$$\begin{array}{r} 32{,}709 \\ +93{,}804 \\ \hline \end{array} \qquad \begin{array}{r} 318 \\ -296 \\ \hline \end{array} \qquad \begin{array}{r} 9722 \\ -8756 \\ \hline \end{array} \qquad \begin{array}{r} 2546 \\ +8369 \\ \hline \end{array}$$

2.

$$\begin{array}{r} 736 \\ \times\ 84 \\ \hline \end{array} \qquad 13\overline{)6903} \qquad \begin{array}{r} 912 \\ \times\ 67 \\ \hline \end{array} \qquad 8\overline{)944}$$

Division of Fractions by Whole Numbers

To divide a fraction by a whole number, multiply by the reciprocal of the whole number. The reciprocal of a whole number is 1 divided by that number. For example, the reciprocal of 3 is $\frac{1}{3}$. Simplify the answer.

Find: $\frac{4}{7} \div 4$

Multiply by the reciprocal of the whole number.	Cancel.	Multiply.
$\frac{4}{7} \div 4 = \frac{4}{7} \times \frac{1}{4}$	$\frac{\overset{1}{4}}{7} \times \frac{1}{\underset{1}{4}}$	$\frac{1 \times 1}{7 \times 1} = \frac{1}{7}$

GUIDED PRACTICE

Write the reciprocal.

	a	b	c	d	e
1.	3 _$\frac{1}{3}$_	9 ____	7 ____	2 ____	16 ____
2.	10 ____	125 ____	36 ____	21 ____	48 ____

Divide. Simplify.

      a   b

3. $\frac{3}{4} \div 3 = \frac{\overset{1}{3}}{4} \times \frac{1}{\underset{1}{3}} = \frac{1 \times 1}{4 \times 1} = \frac{1}{4}$ $\frac{6}{7} \div 3 = \frac{6}{7} \times \frac{1}{3} =$

4. $\frac{5}{6} \div 7 = \frac{5}{6} \times \frac{1}{7} =$ $\frac{8}{15} \div 4 = \frac{8}{15} \times \frac{1}{4} =$

5. $\frac{9}{10} \div 9 = \frac{9}{10} \times \frac{1}{9} =$ $\frac{12}{25} \div 5 = \frac{12}{25} \times \frac{1}{5} =$

6. $\frac{2}{3} \div 6 = \frac{2}{3} \times$ $\frac{4}{5} \div 10 = \frac{4}{5} \times$

7. $\frac{1}{6} \div 2 = \frac{1}{6} \times$ $\frac{4}{9} \div 2 = \frac{4}{9} \times$

8. $\frac{11}{12} \div 7 = \frac{11}{12} \times$ $\frac{15}{16} \div 10 = \frac{15}{16} \times$

PRACTICE

Divide. Simplify.

	a	*b*
1.	$\frac{7}{8} \div 14 =$	$\frac{5}{6} \div 9 =$
2.	$\frac{2}{5} \div 10 =$	$\frac{13}{16} \div 4 =$
3.	$\frac{4}{7} \div 2 =$	$\frac{9}{16} \div 9 =$
4.	$\frac{3}{4} \div 12 =$	$\frac{11}{12} \div 3 =$
5.	$\frac{1}{7} \div 4 =$	$\frac{3}{10} \div 6 =$
6.	$\frac{1}{3} \div 15 =$	$\frac{15}{16} \div 5 =$

MIXED PRACTICE

Find each answer.

	a	*b*	*c*	*d*
1.	8 3 7 +9 0 5	7 9 2 −1 9 9	1 3,7 0 2 +6 9,8 4 9	5 0,0 0 0 −2 9,8 0 6
2.	3 2 9 × 7	4 8 2 × 4	8 7 1 × 2 6	4 2 3 × 9 5
3.	4 1)9 8 4	1 7)9 5 2	8)2 5 6	9 4)1 0 3 4

Division of Whole Numbers by Fractions

To divide a whole number by a fraction, write the whole number as an improper fraction. Multiply by the reciprocal of the second fraction. Simplify the answer.

Find: $6 \div \dfrac{2}{5}$

Write the whole number as an improper fraction.	Multiply by the reciprocal of the second fraction.	Cancel.	Multiply. Simplify.
$6 \div \dfrac{2}{5} = \dfrac{6}{1} \div \dfrac{2}{5}$	$\dfrac{6}{1} \times \dfrac{5}{2}$	$\dfrac{\overset{3}{\cancel{6}}}{1} \times \dfrac{5}{\underset{1}{\cancel{2}}}$	$\dfrac{3 \times 5}{1 \times 1} = \dfrac{15}{1} = 15$

PRACTICE

Divide. Simplify.

a 　　　　　　　　　　　　　　　　　　　　b

1. $1 \div \dfrac{1}{2} = \dfrac{1}{1} \div \dfrac{1}{2} = \dfrac{1}{1} \times \dfrac{2}{1} = \dfrac{1 \times 2}{1 \times 1} = \dfrac{2}{1} = 2$ 　　　$6 \div \dfrac{3}{4} =$

2. $7 \div \dfrac{7}{8} =$ 　　　　　　　　　　　$14 \div \dfrac{7}{8} =$

3. $8 \div \dfrac{2}{5} =$ 　　　　　　　　　　　$9 \div \dfrac{3}{16} =$

4. $15 \div \dfrac{5}{6} =$ 　　　　　　　　　　$2 \div \dfrac{4}{9} =$

5. $21 \div \dfrac{7}{20} =$ 　　　　　　　　　$4 \div \dfrac{1}{12} =$

6. $10 \div \dfrac{5}{18} =$ 　　　　　　　　　$8 \div \dfrac{6}{7} =$

7. $2 \div \dfrac{1}{3} =$ 　　　　　　　　　　$3 \div \dfrac{9}{10} =$

8. $6 \div \dfrac{1}{2} =$ 　　　　　　　　　　$1 \div \dfrac{5}{6} =$

MULTIPLICATION AND DIVISION OF FRACTIONS

Division of Mixed Numbers by Whole Numbers

To divide a mixed number by a whole number, write the mixed number as an improper fraction. Multiply by the reciprocal of the whole number. Simplify the answer.

Find: $2\frac{5}{8} \div 3$

Write the mixed number as an improper fraction.	Multiply by the reciprocal of the whole number.	Cancel.	Multiply.
$2\frac{5}{8} \div 3 = \frac{21}{8} \div 3$	$\frac{21}{8} \times \frac{1}{3}$	$\overset{7}{\cancel{\frac{21}{8}}} \times \frac{1}{\underset{1}{\cancel{3}}}$	$\frac{7 \times 1}{8 \times 1} = \frac{7}{8}$

PRACTICE

Divide. Simplify.

a *b*

1. $3\frac{1}{2} \div 7 = \frac{7}{2} \div 7 = \overset{1}{\cancel{\frac{7}{2}}} \times \frac{1}{\underset{1}{\cancel{7}}} = \frac{1 \times 1}{2 \times 1} = \frac{1}{2}$ $1\frac{2}{3} \div 5 =$

2. $4\frac{1}{2} \div 3 =$ $7\frac{1}{2} \div 10 =$

3. $3\frac{1}{7} \div 2 =$ $4\frac{1}{8} \div 3 =$

4. $3\frac{1}{3} \div 5 =$ $5\frac{4}{7} \div 13 =$

5. $6\frac{2}{3} \div 2 =$ $2\frac{2}{3} \div 4 =$

6. $3\frac{3}{4} \div 12 =$ $1\frac{1}{3} \div 2 =$

7. $3\frac{1}{9} \div 14 =$ $4\frac{2}{5} \div 2 =$

8. $8\frac{1}{4} \div 11 =$ $2\frac{1}{4} \div 6 =$

Division of Mixed Numbers by Fractions

To divide a mixed number by a fraction, write the mixed number as an improper fraction. Multiply by the reciprocal of the second fraction. Simplify the answer.

Find: $6\frac{5}{6} \div \frac{5}{6}$

Write the mixed number as an improper fraction.	Multiply by the reciprocal of the second fraction.	Cancel.	Multiply. Simplify.
$6\frac{5}{6} \div \frac{5}{6} = \frac{41}{6} \div \frac{5}{6}$	$\frac{41}{6} \times \frac{6}{5}$	$\frac{41}{\overset{\;}{6}} \times \frac{\overset{1}{6}}{5}$	$\frac{41 \times 1}{1 \times 5} = \frac{41}{5} = 8\frac{1}{5}$

PRACTICE

Divide. Simplify.

a

b

1. $2\frac{1}{3} \div \frac{1}{6} = \frac{7}{3} \div \frac{1}{6} = \frac{7}{\overset{3}{\underset{1}{\cancel{3}}}} \times \frac{\overset{2}{\cancel{6}}}{1} = \frac{7 \times 2}{1 \times 1} = \frac{14}{1} = 14$ $1\frac{1}{3} \div \frac{2}{3} =$

2. $1\frac{7}{12} \div \frac{3}{4} =$ $3\frac{3}{4} \div \frac{1}{2} =$

3. $5\frac{1}{2} \div \frac{1}{2} =$ $3\frac{1}{10} \div \frac{2}{5} =$

4. $1\frac{1}{2} \div \frac{3}{4} =$ $1\frac{1}{5} \div \frac{4}{5} =$

5. $2\frac{1}{5} \div \frac{4}{5} =$ $2\frac{2}{9} \div \frac{4}{5} =$

6. $1\frac{4}{5} \div \frac{2}{3} =$ $1\frac{4}{7} \div \frac{1}{7} =$

7. $5\frac{1}{4} \div \frac{1}{3} =$ $3\frac{2}{3} \div \frac{11}{12} =$

8. $3\frac{3}{4} \div \frac{5}{8} =$ $5\frac{2}{5} \div \frac{3}{5} =$

MULTIPLICATION AND DIVISION OF FRACTIONS

Division of Mixed Numbers by Mixed Numbers

To divide a mixed number by a mixed number, write the mixed numbers as improper fractions. Multiply by the reciprocal of the second fraction. Simplify the answer.

Find: $2\frac{1}{4} \div 3\frac{3}{8}$

Write the mixed numbers as improper fractions.	Multiply by the reciprocal of the second fraction.	Cancel.	Multiply.
$2\frac{1}{4} \div 3\frac{3}{8} = \frac{9}{4} \div \frac{27}{8}$	$\frac{9}{4} \times \frac{8}{27}$	$\overset{1}{\underset{1}{\cancel{\frac{9}{4}}}} \times \overset{2}{\underset{3}{\cancel{\frac{8}{27}}}}$	$\frac{1 \times 2}{1 \times 3} = \frac{2}{3}$

PRACTICE

Divide. Simplify.

a

1. $4\frac{1}{4} \div 8\frac{1}{2} = \frac{17}{4} \div \frac{17}{2} = \overset{1}{\underset{2}{\cancel{\frac{17}{4}}}} \times \overset{1}{\underset{1}{\cancel{\frac{2}{17}}}} = \frac{1 \times 1}{2 \times 1} = \frac{1}{2}$

2. $5\frac{1}{2} \div 1\frac{1}{2} =$

3. $8\frac{1}{4} \div 2\frac{1}{2} =$

4. $6\frac{2}{3} \div 2\frac{1}{5} =$

5. $8\frac{1}{2} \div 4\frac{1}{4} =$

6. $4\frac{1}{2} \div 1\frac{1}{4} =$

7. $2\frac{1}{2} \div 1\frac{1}{3} =$

8. $6\frac{2}{3} \div 2\frac{1}{4} =$

b

$7\frac{1}{2} \div 4\frac{3}{5} =$

$3\frac{1}{3} \div 1\frac{20}{21} =$

$2\frac{1}{3} \div 3\frac{1}{2} =$

$8\frac{2}{3} \div 1\frac{1}{3} =$

$6\frac{4}{5} \div 1\frac{1}{5} =$

$9\frac{7}{9} \div 1\frac{5}{6} =$

$1\frac{3}{8} \div 3\frac{2}{3} =$

$4\frac{2}{9} \div 1\frac{7}{12} =$

MULTIPLICATION AND DIVISION OF FRACTIONS

Customary Measurement

Some customary units of length are the inch, foot, yard, and mile. The chart shows the relationship of one unit to another.

1 foot (ft)   = 12 inches (in.)
1 yard (yd) = 3 ft
= 36 in.
1 mile (mi) = 1760 yd
= 5280 ft

Some customary units of weight are the ounce, pound, and ton. The chart shows the relationship of one unit to another.

1 pound (lb) = 16 ounces (oz)
1 ton (T) = 2000 pounds

Some customary units of capacity are the cup, pint, quart, and gallon. The chart shows the relationship of one unit to another.

1 pint (pt) = 2 cups (c)
1 quart (qt) = 2 pt
= 4 c
1 gallon (gal) = 4 qt
= 8 pt
= 16 c

PRACTICE

Change each measurement to a larger unit.

	a	b	c
1.	39 in. = _____ ft _____ in.	52 in. = _____ ft _____ in.	18 in. = _____ ft _____ in.
2.	35 oz = _____ lb _____ oz	60 oz = _____ lb _____ oz	21 oz = _____ lb _____ oz
3.	17 pt = _____ qt _____ pt	7 c = _____ pt _____ c	21 qt = _____ gal _____ qt

Change each measurement to a smaller unit.

	a	b	c
4.	6 yd = _____ in.	4 ft = _____ in.	2 mi = _____ ft
5.	8 lb = _____ oz	12 lb = _____ oz	3 T = _____ lb
6.	7 qt 1 c = _____ c	16 pt = _____ c	5 gal = _____ pt

PROBLEM SOLVING

Applications

Solve.

1. In a $4\frac{1}{2}$-mile relay race, there were 9 runners. What part of a mile did each run?

Answer _____

2. Sue bought 54 feet of lumber for building window frames. Each window frame requires $13\frac{1}{2}$ feet of lumber. How many window frames can she make?

Answer _____

3. Alfredo was planting tomato plants in a row $6\frac{2}{3}$ yards long. He read that plants should be $\frac{2}{3}$ yard apart. How many plants were in the row?

Answer _____

4. The Sernas bought material to be used for making curtains. There are $24\frac{1}{2}$ yards of material. Each window requires $3\frac{1}{2}$ yards. How many sets of curtains can the Sernas make?

Answer _____

5. ACME Contractors bought a 12-pound carton of glue. The glue was in $\frac{1}{8}$ pound tubes. How many tubes were there in the carton?

Answer _____

6. A load of cement weighing $\frac{3}{4}$ ton was divided into bags, each weighing $\frac{1}{20}$ ton. How many bags did the load contain?

Answer _____

7. Jane was leveling a washing machine. One corner needed to be raised $\frac{3}{4}$ inch. She had pieces of wood $\frac{1}{16}$ inch thick. How many pieces of wood would she need to put under the corner?

Answer _____

8. Fred brought 6 pounds of potato salad to the party. How many $\frac{1}{4}$ pound servings were there?

Answer _____

Find a Pattern

The answer to a problem may be found by recognizing a pattern. Read the problem carefully. Write the pattern and determine how the numbers are related. Find the rule that makes the pattern. Then solve the problem.

EXAMPLE 1

Read the problem.

What is the next number in this number pattern?
7, 21, 63, 189, . . .

Determine the relationship.

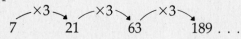

Write the rule.

Multiply by 3.

Solve the problem.

$189 \times 3 = 567$. The next number is 567.

EXAMPLE 2

Read the problem.

The height of a tomato plant has increased over the last two weeks. During Week 1, the plant grew from $2\frac{1}{4}$ feet to $2\frac{7}{8}$ feet. During Week 2, the plant grew to $3\frac{1}{2}$ feet. If the plant continues to grow at this same rate, how tall will the plant be at the end of Week 3?

Write the pattern.

The number pattern was not given to you. A table will help you write the pattern.

Week	0	1	2	3
Height in feet	$2\frac{1}{4}$	$2\frac{7}{8}$	$3\frac{1}{2}$	?

Determine the relationship.

$$2\frac{1}{4} \xrightarrow{+\frac{5}{8}} 2\frac{7}{8} \xrightarrow{+\frac{5}{8}} 3\frac{1}{2} \ldots$$

Write the rule.

Add $\frac{5}{8}$ foot for each week.

Solve the problem.

$$3\frac{1}{2} + \frac{5}{8} = 3\frac{4}{8} + \frac{5}{8} = 3\frac{9}{8} = 4\frac{1}{8}$$

The plant will be $4\frac{1}{8}$ feet tall at the end of Week 3.

Solve by finding the pattern. Write the rule. Then answer the question.

1. What is the next number in the pattern?
   7, 9, 11, . . .

   Rule _____

   Answer _____

2. What are the next two numbers in this pattern?
   48, 44, 40, . . .

   Rule _____

   Answer _____

3. What is the next number in this pattern?
   $8\frac{2}{3}$, 10, $11\frac{1}{3}$, $12\frac{2}{3}$, . . .

   Rule _____

   Answer _____

4. What are the next two numbers in this pattern?
   5, 15, 45, 135, . . .

   Rule _____

   Answer _____

5. What is the missing number in this pattern?
   1, 2, 4, _____, 16, 32, . . .

   Rule _____

   Answer _____

6. What is the missing number in this pattern?
   1280, 320, _____, 20

   Rule _____

   Answer _____

7. Mr. Rivera's salary increased last year from $18,260 to $19,040. His present salary is $19,820. At this rate, what will his salary be at the end of next year?

   Rule _____

   Answer _____

8. The value of stock in the Acme Corporation has gone from $58\frac{1}{8}$ per share to $54\frac{3}{4}$ to $51\frac{3}{8}$ over the past 3 months. How much will the stock be worth next month if the trend continues?

   Rule _____

   Answer _____

PROBLEM SOLVING

Applications

Solve.

1. How many cups of juice can be served from a container that holds $2\frac{1}{2}$ gallons?

   Answer _____

2. A crew has paved $5\frac{1}{2}$ miles of a 7-mile road. How many miles do they have left to pave?

   Answer _____

3. Mark needs 59 inches of tape for boxes. He has $2\frac{1}{2}$ feet of tape on one roll and $3\frac{1}{2}$ feet of tape on another roll. How many inches does he have to spare?

   Answer _____

4. How many yards of weather stripping are needed to go around a window that measures 20 feet around its border?

   Answer _____

5. A truck is hauling 4 cars that each weigh $2\frac{1}{2}$ tons. How many tons does the truck's load weigh?

   Answer _____

6. Jeff cut $1\frac{1}{2}$ feet from a 6-foot piece of lumber. How many feet are left of the original piece?

   Answer _____

7. Martha used $2\frac{1}{2}$ cups of milk for a cream soup and $4\frac{1}{2}$ cups of milk for a dessert. How many cups of milk did she use in all?

   Answer _____

8. Kim used 3 quarts and 1 pint of chicken broth and 2 pints of beef broth in his soup. How many pints did he use in all?

   Answer _____

MULTIPLICATION AND DIVISION OF FRACTIONS
Unit 4 Review

Multiply or divide. Simplify.

a	*b*

1. $\dfrac{2}{3} \times \dfrac{1}{5} =$ $\qquad$ $\dfrac{7}{8} \div 7 =$

2. $1\dfrac{2}{5} \times \dfrac{5}{12} =$ $\qquad$ $\dfrac{3}{5} \times \dfrac{5}{9} =$

3. $4\dfrac{4}{9} \times \dfrac{9}{10} =$ $\qquad$ $1\dfrac{2}{5} \div \dfrac{7}{10} =$

4. $2\dfrac{6}{7} \div 3\dfrac{1}{14} =$ $\qquad$ $\dfrac{1}{10} \div \dfrac{1}{20} =$

5. $5 \div \dfrac{3}{4} =$ $\qquad$ $5 \times \dfrac{3}{4} =$

6. $6\dfrac{1}{8} \times \dfrac{3}{7} =$ $\qquad$ $4\dfrac{1}{4} \div \dfrac{1}{8} =$

7. $15 \div \dfrac{7}{10} =$ $\qquad$ $\dfrac{4}{15} \div \dfrac{1}{3} =$

8. $2\dfrac{7}{10} \times 2\dfrac{1}{5} =$ $\qquad$ $\dfrac{7}{8} \div \dfrac{3}{4} =$

9. $9 \div \dfrac{2}{3} =$ $\qquad$ $9 \div \dfrac{3}{5} =$

10. $8\dfrac{1}{8} \times 1\dfrac{3}{5} =$ $\qquad$ $\dfrac{1}{10} \div \dfrac{1}{10} =$

Solve.

11. Tonia was using metal spacers to separate two steel plates. Each spacer was $\dfrac{3}{16}$ inch thick. How many spacers did she need to separate the plates $1\dfrac{1}{2}$ inches?

12. From Chicago to St. Louis is 273 miles. At an average speed of $55\dfrac{1}{2}$ miles per hour, how long did it take to drive from Chicago to St. Louis?

Answer _____

Answer _____

Reading and Writing Decimals

To read a decimal, read as a whole number. Then name the place value of the last digit.

Read and write 0.53 as fifty-three hundredths.

To read a decimal that has a whole number part,

- read the whole number part.
- read the decimal point as "and".
- read the decimal part as a whole number and then name the place value of the last digit.

Read and write 23.705 as twenty-three and seven hundred five thousandths.

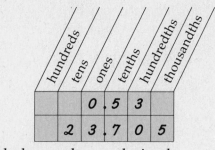

← whole number . decimal →

PRACTICE

Write as a decimal.

a   *b*

1. three tenths ____0.3____   twenty-five hundredths _____

2. fifteen thousandths _____   one and five tenths _____

3. ten and four hundredths _____   six and sixty-six thousandths _____

4. one hundred seventy-five thousandths _____

Write each decimal in words.

5. 0.005 ____*five thousandths*_____

6. 39.374 _____

7. $1.23 _____

8. $14.08 _____

9. 0.06 _____

Write each money amount with a dollar sign and a decimal point.

a   *b*   *c*

10. nine dollars ____$9.00____   ninety cents _____   nine cents _____

11. sixty-six cents _____   eleven cents _____   forty-two dollars _____

12. one hundred ten dollars and seventy-four cents _____

13. two thousand, five dollars and three cents _____

14. one dollar and nineteen cents _____

WORKING WITH DECIMALS
Compare and Order Decimals

To compare two decimal numbers, begin at the left. Compare the digits in each place.

The symbol < means "is less than." *4.2 < 4.6*

The symbol > means "is greater than." *2.7 > 2.3*

The symbol = means "is equal to." *3.4 = 3.40*

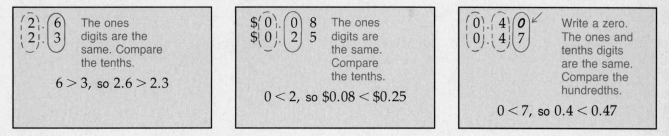

Compare: 2.6 and 2.3

| 2 . 6 | The ones |
| 2 . 3 | digits are the same. Compare the tenths. |

6 > 3, so 2.6 > 2.3

Compare: $0.08 and $0.25

| $ 0 . 0 8 | The ones |
| $ 0 . 2 5 | digits are the same. Compare the tenths. |

0 < 2, so $0.08 < $0.25

Compare: 0.4 and 0.47

| 0 . 4 0 | Write a zero. |
| 0 . 4 7 | The ones and tenths digits are the same. Compare the hundredths. |

0 < 7, so 0.4 < 0.47

PRACTICE

Compare. Write <, >, or =.

	a	*b*	*c*
1.	0.3 __<__ 0.32 0 . 3 0 0 . 3 2	0.035 _____ 0.35	0.5 _____ 0.500
2.	0.125 _____ 13	0.15 _____ 0.115	3.5 _____ 3.50
3.	0.620 _____ 0.62	0.35 _____ 0.350	0.26 _____ 0.3
4.	5.56 _____ 5.561	0.95 _____ 0.905	2.65 _____ 2.6
5.	0.25 _____ 0.3	0.6 _____ 0.65	3.008 _____ 3.080
6.	4.50 _____ 4.500	0.78 _____ 0.789	2.001 _____ 2.1

Write in order from least to greatest.

	a	*b*
7.	67.5 0.675 60.80 _____	7.026 7.260 7.230 _____
8.	1.025 1.20 1.1 _____	0.34 0.034 0.304 _____

109

Fraction and Decimal Equivalents

Sometimes you will need to either change a decimal to a fraction or a fraction to a decimal.

To write a decimal as a fraction, identify the value of the last place in the decimal. Use this place value to write the denominator. Simplify if possible.

To write a fraction that has a denominator of 10, 100, or 1000 as decimal, write the digits from the numerator. Write the decimal point. Notice that the number of zeros is the same as the number of decimal places in the decimal.

Decimal		Fraction or Mixed Number
0.9	=	$\frac{9}{10}$
0.01	=	$\frac{1}{100}$
0.045	=	$\frac{45}{1000} = \frac{9}{200}$
1.74	=	$\frac{174}{100} = 1\frac{74}{100} = 1\frac{37}{50}$

Fraction or Mixed Number		Decimal
$\frac{3}{10}$	=	0.3
$\frac{15}{100}$	=	0.15
$\frac{6}{1000}$	=	0.006
$\frac{59}{10}$ or $5\frac{9}{10}$	=	5.9

PRACTICE

Write each decimal as a fraction.

	a	b	c	d
1.	0.4 $\frac{4}{10}$	0.6 _____	0.08 _____	0.002 _____
2.	0.21 _____	0.083 _____	0.901 _____	0.018 _____

Write each decimal as a mixed number.

	a	b	c	d
3.	4.5 $4\frac{5}{10}$	1.62 _____	10.1 _____	1.275 _____
4.	9.07 _____	38.24 _____	5.46 _____	13.8 _____

Write each fraction as a decimal.

	a	b	c	d
5.	$\frac{1}{10}$ 0.1	$\frac{2}{10}$ _____	$\frac{5}{10}$ _____	$\frac{7}{10}$ _____
6.	$\frac{6}{100}$ _____	$\frac{80}{100}$ _____	$\frac{52}{1000}$ _____	$\frac{416}{1000}$ _____
7.	$\frac{56}{10}$ _____	$\frac{31}{10}$ _____	$\frac{76}{10}$ _____	$\frac{65}{100}$ _____
8.	$\frac{103}{100}$ _____	$\frac{509}{100}$ _____	$\frac{1643}{1000}$ _____	$\frac{2051}{1000}$ _____

Fraction and Decimal Equivalents

Not all fractions can be changed to decimal form easily. To write fractions that have denominators other than 10, 100, or 1000 as decimals, first write an equivalent fraction that has a denominator of 10, 100, or 1000. Then write the equivalent fraction as a decimal.

Remember, not all fractions have simple decimal equivalents.

Examples: $\frac{2}{3} = 0.666\ldots$ $\frac{5}{6} = 0.833\ldots$

Write $\frac{1}{5}$ as a decimal.

Write $\frac{1}{5}$ with 10 as the denominator.	Write the fraction as a decimal.
$\frac{1}{5} = \frac{1 \times 2}{5 \times 2} = \frac{2}{10}$	$= 0.2$

Write $2\frac{3}{4}$ as a decimal.

Write $2\frac{3}{4}$ as an improper fraction.	Write the new fraction with 100 as the denominator.	Write the fraction as a decimal.
$2\frac{3}{4} = \frac{11}{4}$	$\frac{11}{4} = \frac{11 \times 25}{4 \times 25} = \frac{275}{100}$	$= 2.75$

PRACTICE ──────────────

First find an equivalent fraction that has a denominator of 10, 100, or 1000. Then write each fraction as a decimal.

	a	*b*	*c*
1.	$\frac{1}{8} = \frac{1 \times 125}{8 \times 125} = \frac{125}{1000} = 0.125$	$\frac{2}{5} =$	$\frac{3}{4} =$
2.	$\frac{4}{5} =$	$\frac{17}{50} =$	$\frac{11}{25} =$
3.	$\frac{7}{20} =$	$\frac{8}{25} =$	$\frac{3}{8} =$
4.	$\frac{13}{2} = \frac{13 \times 5}{2 \times 5} = \frac{65}{10} = 6.5$	$\frac{43}{20} =$	$\frac{37}{5} =$
5.	$\frac{25}{4} =$	$\frac{69}{50} =$	$\frac{39}{25} =$

Write each mixed number as a decimal.

	a	*b*
6.	$1\frac{9}{20} = \frac{29}{20} = \frac{29 \times 5}{20 \times 5} = \frac{145}{100} = 1.45$	$2\frac{21}{25} =$
7.	$6\frac{3}{4} =$	$13\frac{1}{5} =$
8.	$19\frac{1}{2} =$	$4\frac{7}{8} =$

Addition of Decimals

To add decimals, line up the decimal points. Write zeros as needed. Then add as with whole numbers. Be sure to write a decimal point in the sum.

Find: 4.6 + 7.32

Write a zero.		Add the hundredths.		Add the tenths. Write a decimal point in the sum.		Add the ones.	

T	O	Ts	Hs
	4	.6	0
+ 7		.3	2

T	O	Ts	Hs
	4	.6	0
+ 7		.3	2
			2

T	O	Ts	Hs
	4	.6	0
+ 7		.3	2
		.9	2

T	O	Ts	Hs
	4	.6	0
+ 7		.3	2
1	1	.9	2

GUIDED PRACTICE

Add. Write zeros as needed.

a

1.

O	Ts	Hs
	1	
$3	.1	9
+ 2	.2	2
$5	.4	1

2.

O	Ts	Hs	Ths
0	.3	*0*	*0*
+0	.0	0	6
0	.3	0	6

3.

T	O	Ts	Hs
1		*1*	
$	2	.1	0
	4	.0	8
+	5	.2	5
$1	1	.4	3

4.

T	O	Ts	Hs	Ths
1	*2*	*2*		
	4	.0	9	*0*
3	3	.9	8	4
1	0	.4	*0*	*0*
+2	3	.7	3	*0*
7	2	.2	0	4

b

1.

O	Ts	Hs
$0	.0	2
+ 0	.5	7

2.

T	O	Ts	Hs	Ths
2	2	.1	3	
+	7	.0	9	8

3.

T	O	Ts	Hs	Ths
1	4	.0	0	5
1	6	.1	9	3
+	7	.3	2	7

4.

T	O	Ts	Hs	Ths
9		.4	5	1
0		.0	0	7
7		.3		
+	6	.5	3	6

c

1.

T	O	Ts	Hs
$1	4	.9	0
+	1	.9	6

2.

H	T	O	Ts
2	3	8	
+	7	7	.3

3.

O	Ts	Hs	Ths
0	.3	5	6
0	.4	3	0
+0	.8	1	7

4.

H	T	O	Ts	Hs
$2	3	1	.2	4
	6	1	.6	
1	4	2	.0	5
+ 1	0	0	.1	1

d

1.

T	O	Ts	Hs
$4	2	.0	8
+	0	.1	6

2.

T	O	Ts	Hs
2	4	.1	5
+3	0	.1	

3.

T	O	Ts	Hs	Ths
1	4	.0	0	5
	0	.0	4	3
+	2	.6	8	9

4.

T	O	Ts	Hs	Ths
5	6	.1	1	2
	4	.2	2	
2	4	.7	3	6
+1	4			

Add. Write zeros as needed.

	a	b	c	d
1.	5.9 +3.6 2	7.0 8 +3.2 6 5	1 4.0 7 6 + 8.4 6	1 7.0 5 + 3.3 5 1
2.	2 8.0 0 9 4.6 5 + 6.0 0 3	7 7.0 1 6 4.5 7 + 0.6 4 7	8 4.7 0.4 0 3 + 3.0 8	8.2 3 3 6.5 + 0.0 0 9
3.	$1 2.2 0 2.1 0 4.0 8 + 5.2 5	8.0 5 0 1 4.0 0 5 1 6.1 + 7.3 2	0.2 2 0.3 5 6 0.4 3 9 +0.8	6.1 2 6 1 4.0 0 5 0.0 4 + 2.6
4.	2.0 0 5 0.1 5 0.6 +3	0.0 0 9 0.1 4 0.6 +2.1 0	0.1 4 0.0 6 0 0.1 7 +8.5	0.7 0 2 0.0 0 5 0.0 3 4 +7.1 6

Line up the digits. Then add. Write zeros as needed.

a   b

5. 9 + 3.4 + 0.7 = _____ $6.54 + $10 + $8.35 = _____

9.0↙
3.4
+0.7

6. 6.9 + 12.7 + 38.6 = _____   37.5 + 5.3 + 8 + 3.273 = _____

MIXED PRACTICE

Multiply or divide.

	a	b	c	d
1.	2 3 4 × 1 5	3 1 5 × 2 8	3 5)7 7 0	1 8)4 0 5 0

113

Subtraction of Decimals

To subtract decimals, line up the decimal points. Write zeros as needed. Then subtract as with whole numbers. Be sure to write a decimal point in your answer.

Find: 34.3 − 17.94

Write a zero. Regroup to subtract the hundredths.	Regroup to subtract the tenths. Write a decimal point in the difference.	Regroup to subtract the ones.	Subtract the tens.

T	O	Ts	Hs
		2	10
3	4	.3̶	0̶ ✓
−1	7	.9	4
			6

T	O	Ts	Hs
		12	
	3	2̶	10
3	4̶	.3̶	0̶
−1	7	.9	4
		.3	6

T	O	Ts	Hs
	13	12	
2	8̶	2̶	10
3̶	4	.3̶	0̶
−1	7	.9	4
	6	.3	6

T	O	Ts	Hs
	13	12	
2	3̶	2̶	10
3̶	4̶	.3̶	0̶
−1	7	.9	4
1	6	.3	6

GUIDED PRACTICE

Subtract. Write zeros as needed.

1.

a

O	Ts	Hs	Ths
5	.4	9	8
−2	.3	6	2
3	.1	3	6

b

O	Ts	Hs
7	.5	4
−6	.3	8

c

T	O	Ts	Hs	Ths
1	5	.0	6	5
−		9	.4	6	6

d

T	O	Ts	Hs
$4	7	.6	2
− 2	3	.8	5

2.

a

O	Ts	Hs	Ths
6	10	4	10
7̶	.0̶	5̶	0̶ ✓
−3	.1	3	5
3	.9	1	5

b

T	O	Ts	Hs
1	2	.5	0 ✓
−		7	.7	5

c

O	Ts	Hs
8	.1	4 ✓
−6	.1	0

d

O	Ts	Hs	Ths
2	.8	7	7
−0	.9	8	0

3.

a

O	Ts	Hs	Ths
7	.5	8	9
−3	.3	4	7

b

T	O	Ts	Hs
2	4	.3	6
−		7	.1	6

c

O	Ts	Hs	Ths
8	.1	1	3
−7	.3	0	5

d

T	O	Ts	Hs
$2	1	.0	9
− 1	6	.9	5

4.

a

O	Ts	Hs	Ths
7	10	8	10
8̶	.0̶	0̶	0̶ ✓
−4	.2	5	6
3	.8	3	4

b

O	Ts	Hs	Ths
1	.5	2	0 ✓
−0	.4	0	8

c

T	O	Ts	Hs
7	5	.1	6 ✓
−5	2	.8	0

d

T	O	Ts	Hs
4	0	.3	3 ✓
−2	9	.7	0

Subtract. Write zeros as needed.

	a	b	c	d
1.	8.3 2 5 −3.2 0 3	7.2 7 8 −5.1 2	9.0 6 8 −7.0 5 4	1 0.3 9 9 −1 0.2 3 9
2.	$3 4.9 5 − 2 7.9 9	$9 2.0 0 − 6 7.5 0	$9 4.7 8 − 1 5.0 0	$1 1,5 3 2.3 0 − 2,5 0 0.0 0
3.	6.2 −4.5 7 5	1.9 −0.6 7 4	1.3 5 4 −0.2 6 5	9 2.1 5 −8 4.7
4.	8.0 9 −4.2 5 6	9 4.7 8 −1 5	1 9.0 0 5 −1 4.5	1.5 2 −0.4 0 8

Line up the digits. Then subtract. Write zeros as needed.

	a	b	c
5.	7.05 − 3.035 _____ 7.050 ↙ −3.035	8.14 − 6.1 _____	75.06 − 52.8 _____
6.	92.15 − 84.7 _____	12.5 − 7.75 _____	1.354 − 0.265 _____

➡ **MIXED PRACTICE**

Write as a decimal.

	a	b
1.	six thousandths _____	twelve hundredths _____
2.	twenty thousandths _____	four hundredths _____
3.	nine tenths _____	forty thousandths _____
4.	three and three tenths _____	six and eight thousandths _____

Compare. Write <, >, or =.

	a	b	c	d
5.	2.10 ___ 2.1	0.456 ___ 4.56	7.09 ___ 7.009	8.110 ___ 8.11
6.	0.078 ___ 0.08	3.405 ___ 3.4	19.230 ___ 19.23	0.89 ___ 0.90

PROBLEM-SOLVING STRATEGY

Use Estimation

Many problems can be solved by estimation. Often, you do not need an exact answer to solve a problem. An estimate is found by rounding some or all of the numbers and then doing mental math.

Read the problem.

> Green Things Plant Shop was having a sale. Tulips were on sale for $0.45 each. Cacti were on sale for $0.79 each. Rose bushes were on sale for $2.99 each. Joe has $5.00. If he buys one rose bush, can he also buy two cactus plants?

Identify the important facts.

> Joe has $5.00.
>
> One rose bush costs $2.99.
>
> One cactus plant costs $0.79.

Round.

> Round $2.99 to the nearest dollar.
> > $2.99 rounds to $3.
>
> Round $0.79 to the nearest tenth of a dollar (dime).
> > $0.79 rounds to $0.80.

Solve the problem.

> $3.00 + 2 ($0.80) = $3.00 + $1.60 = $4.60
>
> Joe has enough money to buy one rose bush and two cactus plants.

Use estimation to solve each problem.

---

1. Polly wants to buy 2 pounds of T-bone steak at $3.79 per pound. About how much money does she need?

    Answer _____

2. Giorgio has $20.00. Can he buy 3 pounds of sirloin steak at $2.39 per pound and 4 pounds of round steak at $1.99 per pound?

    Answer _____

---

3. James bought 3 pounds of round steak at $1.99 per pound and 4 pounds of chuck steak at $1.69 per pound. About how much money did he get back from his 20-dollar bill?

    Answer _____

4. T-bone steak costs $3.79 per pound. About how many pounds of T-bone steak can you buy for $30?

    Answer _____

Use estimation to solve each problem.

5. A store owner purchased 34 videotapes at $8.85. About how much money did she spend in all?

Answer _____

6. One hundred quarters weigh about 1.25 pounds. About how many quarters are there in 20 pounds?

Answer _____

7. Pedro bought two birthday presents for $7.65 each. He gave the clerk $20.00. About how much money did he get back?

Answer _____

8. Charlie wanted to buy a pair of socks for $3.99, a belt for $12.39, and a tie for $7.88. Would $20 be enough money for his purchase?

Answer _____

9. Babba the elephant weighed 14,256 pounds. This was 1977 pounds more than her sister Lissa's weight. About how much did Lissa weigh?

Answer _____

10. Gretchen saves $36 each month. About how much money will she save in two years?

Answer _____

11. Angela earns $6.35 every week doing jobs around the house. About how much does Angela earn in one year? (1 year = 52 weeks)

Answer _____

12. Byron wants to estimate his electric bill for the coming year. His current monthly bill is $64.78. About how much will Byron pay for electricity in one year?

Answer _____

PROBLEM SOLVING

Applications

Solve.

1. Connie was comparing shoe prices. She found one pair that cost $26.95. She found a similar pair that cost $1.92 more. How much did the second pair cost?

Answer _____

2. Rosa delivers packages. Rosa drove to Sandoval, a distance of 8.2 kilometers. Then she drove 16.5 kilometers to Richview. From there, she drove 14.9 kilometers to Mount Vernon. Finally she drove 39.9 kilometers to Marion. How far did Rosa drive in all?

Answer _____

3. Jan is painting her kitchen. She bought a gallon of paint for $11.47, a brush for $2.49, masking tape for $0.59, and a package of sandpaper for $1.77. How much did Jan pay for the painting supplies?

Answer _____

4. From New Orleans to El Paso it is 1975.7 kilometers. Tucson is 500.5 kilometers farther. How far is it from New Orleans to Tucson through El Paso?

Answer _____

This table shows the distance in kilometers by rail from Chicago to New Orleans and intermediate cities.

0.0	Chicago, IL	850.3	Memphis, TN
204.1	Champaign, IL	1010.2	Grenada, MS
580.9	Cairo, IL	1173.7	Jackson, MS
652.9	Fulton, KY	1399.7	Hammond, LA
786.9	Covington, KY	1485.6	New Orleans, LA

5. How far is it from Chicago to Cairo?

Answer _____

6. How far is it from Chicago to New Orleans?

Answer _____

7. How far is Memphis from New Orleans?

Answer _____

8. How far is it from Covington to New Orleans?

Answer _____

WORKING WITH DECIMALS

Estimation of Decimal Sums and Differences

To estimate decimal sums, first round the decimals to the same place. Then add the rounded numbers.

To estimate decimal differences, first round the decimals to the same place. Then subtract the rounded numbers.

Estimate: $7.69 + $4.19

Round each decimal to the nearest dollar. Add.

$$
\begin{array}{r}
\$7.69 \rightarrow \quad \$\,8 \\
+\ 4.19 \rightarrow +\quad 4 \\
\hline
\$12
\end{array}
$$

Estimate: 10.34 − 6.78

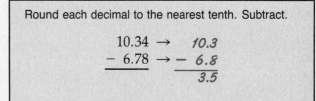

Round each decimal to the nearest tenth. Subtract.

$$
\begin{array}{r}
10.34 \rightarrow \quad 10.3 \\
-\ 6.78 \rightarrow -\quad 6.8 \\
\hline
3.5
\end{array}
$$

PRACTICE

Estimate the sum or difference by rounding to the nearest one.

	a	*b*	*c*	*d*

1.
- a: $7.65 → $ 8 ; + 5.33 → + 5 ; $13
- b: $10.45 → ; + 23.56 →
- c: $ 99.90 → ; + 121.25 →
- d: $26.76 → ; + 1.98 →

2.
- a: $99.76 → $100 ; − 20.30 → − 20 ; $ 80
- b: $10.45 → ; − 9.23 →
- c: $9.87 → ; − 5.57 →
- d: $13.45 → ; − 12.89 →

3.
- a: $8.54 → ; − 4.50 →
- b: $30.68 → ; + 24.10 →
- c: $23.05 → ; +101.86 →
- d: $29.70 → ; − 8.29 →

Estimate the sum or difference by rounding to the nearest tenth.

	a	*b*	*c*

4. 12.35 − 2.17 ; 5.08 + 3.07 ; 19.18 − 9.28

$$
\begin{array}{r}
12.35 \rightarrow \quad 12.4 \\
-\ 2.17 \rightarrow -\quad 2.2 \\
\hline
10.2
\end{array}
$$

5. 10.67 + 9.33 ; 4.75 − 0.66 ; 300.31 + 25.32

6. 25.04 + 4.86 ; 30.28 − 9.83 ; 175.37 + 24.62

119

Practice Adding and Subtracting Decimals

Estimate the answer. Then add or subtract. Use your estimate to check your answer.

PRACTICE

Add.

	a	b	c	d
1.	0.4 0.2 + 0.3	0.0 2 0.0 5 + 0.0 1	1.0 8 2.0 2 + 0.4 5	4.5 7 2.9 3 + 4.8 7
2.	3.0 6 4.0 9 2.0 8 + 1.0 1	0.0 8 0.0 3 0.0 2 + 0.0 6	0.1 5 0.0 8 0.4 3 + 0.1 7	0.2 5 0.6 0.1 0 + 0.0 5
3.	2.8 6 0.7 0.1 2 + 0.0 8 1	4 0.0 5 1.6 + 0.0 0 3	2.0 3 1 4.1 7 5 3.0 9 8 + 1.0 0 6	1 5 0.4 8 2 5.9 8 1 6.5 0 + 2 5 0.0 0

Subtract.

	a	b	c	d
4.	0.8 − 0.2	1 4.6 − 5.4	0.6 7 − 0.4 8	0.8 6 − 0.7 9
5.	0.6 7 9 − 0.3 9 8	1 5.8 − 3.9	6.5 0 4 − 2.8	7.8 − 1.2 6
6.	0.6 5 9 3 − 0.4 2 7 1	1.8 5 3 1 − 0.9 2 4 8	8 4.3 5 − 3 6.9 5	8.0 0 0 − 1.7 4 2

Line up the decimals. Then find the answer.

	a	b	c

7. 1.24 + 0.078 + 2.9 = _____   18.4957 − 2.36 = _____   324.6 − 75.908 = _____

WORKING WITH DECIMALS

More Practice Adding and Subtracting Decimals

After you find an answer, check with a calculator.

Add.

	a	*b*	*c*	*d*
1.	1.4 0.7 0.2 9 + 2.4 5 6	8 0.5 3 2.1 + 0.6 8	7.0 5 3 0.9 6 8.5 2 4 + 1.8	1.3 8 1 6.4 8 9.2 7 + 0.8 4
2.	8.3 4 0 7 0.0 0 3 8 1.1 5 3 + 7.4 6 1 9	3.9 7 0.9 5 8.4 9 + 3.1 6	2 2.9 5 5.5 0 0.9 8 + 1 6.4 9	3 4 9.9 5 6.5 0 2 4.0 0 + 9.9 9

Line up the decimals. Then add.

           a   *b*

3. $3.3 + 0.07 + 6 + 2.63 + 0.174 =$ _____ $15.4 + 2.185 + 0.66 + 21.009 =$ _____

4. $0.74 + 1.6 + 0.99 + 4.88 + 0.04 =$ _____ $3.42 + 15.98 + 25 + 12.45 =$ _____

Subtract.

	a	*b*	*c*	*d*
5.	0.4 5 − 0.2 1 6	2 3.8 7 − 2.1	4.2 − 0.3 7 2	3.5 6 − 0.8
6.	1 2 − 0.7 4 3	8.7 6 4 5 − 3	7.0 0 0 8 − 2.5	5 6 − 2.0 0 3

Line up the decimals. Then subtract.

           a   *b*   *c*

7. $12.54 - 1.054 =$ _____ $4 - 0.875 =$ _____ $15.42 - 9 =$ _____

8. $27 - 0.0067 =$ _____ $0.03 - 0.0034 =$ _____ $100 - 84.53 =$ _____

WORKING WITH DECIMALS

Multiplying Decimals by Whole Numbers

To multiply decimals by whole numbers, multiply as if you were multiplying whole numbers. Count the number of decimal places to the right of the decimal point in the numbers you have multiplied. The product will have the same number of decimal places. Place the decimal point in the product.

Remember, sometimes you might need to write a zero in the product in order to place the decimal point correctly.

Find: 13 × 6.4

Multiply. Write the decimal point in the product.

```
    6.4      1 decimal place
   ×1 3     +0 decimal places
  ────────  ─────────────────
   19 2
   64
  ────────
   83.2      1 decimal place
```

Find: 0.018 × 5

Multiply. Write the decimal point in the product.

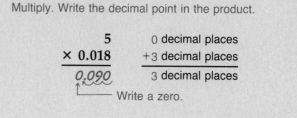

```
      5      0 decimal places
  × 0.018   +3 decimal places
  ────────  ─────────────────
  0.090      3 decimal places
         └── Write a zero.
```

GUIDED PRACTICE

Multiply. Write zeros as needed.

a

1.
```
  ²
  1 7
 ×0.3     __1__ place
 ─────
  5.1
```

2.
```
   2
 ×0.2     __1__ place
 ─────
  0.4
```

3.
```
       ¹
 0.0 0 3    __3__ places
 ×     4
 ───────
 0.0 1 2
```

4.
```
    3.8 4    __2__ places
 ×    3 5
 ─────────
    1 9 2 0
  1 1 5 2 0
 ─────────
  1 3 4.4 0
```

5.
```
   3 2 1
 ×1.4 6    _____ places
```

b

1.
```
   2 4
 ×0.0 9    _____ places
 ───────
```

2.
```
 0.1 0 5    _____ places
 ×     8
 ───────
```

3.
```
     2
 ×0.0 3    _____ places
 ───────
```

4.
```
 0.1 3 5    _____ places
 ×     4 4
 ─────────
```

5.
```
   4.5 3    _____ places
 ×5 7 9
 ───────
```

c

1.
```
 0.7 0 7    _____ places
 ×     2
 ───────
```

2.
```
 0.1 9    _____ places
 ×    5
 ───────
```

3.
```
 0.0 0 6    _____ places
 ×     9
 ───────
```

4.
```
   9.2 8    _____ places
 ×2 3 0
 ───────
```

5.
```
 0.6 9 2    _____ places
 ×   1 6 8
 ─────────
```

Multiply. Write zeros as needed.

	a	b	c	d
1.	0.862 × 2	0.084 × 3	1.63 × 6	2.34 × 5
2.	13.6 × 3	28.52 × 4	1.3 ×13	26 ×1.3
3.	8.2 ×12	62 ×0.35	70 ×5.0	0.90 × 55
4.	3.07 × 25	0.048 × 82	234 ×0.059	707 ×0.690

Line up the digits. Multiply. Write zeros as needed.

	a	b	c
5.	15.2 × 6 = _____	0.908 × 31 = _____	7.85 × 109 = _____

15.2
× 6
—————

➡ MIXED PRACTICE ▬▬▬▬▬▬▬▬▬▬▬▬▬▬▬▬▬▬▬▬▬▬▬▬▬▬
Divide.

	a	b	c	d
1.	23)4577	18)3636	74)9088	69)4815
2.	12)1234	25)1930	50)8040	85)9696

Multiplying Decimals by Decimals

To multiply decimals by decimals, multiply as if you were multiplying whole numbers. Place the decimal point in the product by counting the number of decimal places to the right of the decimal point in both numbers. The product will have the same number of decimal places. Write zeros as needed.

Find: 0.48 × 13.7

Multiply. Write the decimal point in the product.

$$
\begin{array}{rl}
1\,3.7 & \textit{1} \text{ place} \\
\times 0.4\,8 & +\textit{2} \text{ places} \\
\hline
1\,0\,9\,6 & \\
5\,4\,8 & \\
\hline
6.5\,7\,6 & \textit{3} \text{ places}
\end{array}
$$

Find: 0.008 × 0.137

Multiply. Write the decimal point in the product.

$$
\begin{array}{rl}
0.1\,3\,7 & \textit{3} \text{ places} \\
\times 0.0\,0\,8 & +\textit{3} \text{ places} \\
\hline
0.0\,0\,1\,0\,9\,6 & 6 \text{ places}
\end{array}
$$

Write 2 zeros.

GUIDED PRACTICE

Multiply. Write zeros as needed.

	a		*b*		*c*	
1.	²1.6 ×0.4	*1* place + *1* place	5.3 ×0.0 9	place + places	0.7 6 × 0.5	places + place
	0.6 4	2 places		places		places
2.	¹0.1 2 × 0.6	*2* places + *1* place	0.0 9 × 0.3	places + place	0.0 0 2 × 0.4	places + place
	0.0 7 2	3 places		places		places
3.	0.1 8 4 × 0.0 7	places + places	2.0 4 × 0.2	places + place	5.1 9 ×0.0 3	places + places
		places		places		places
4.	2.5 ×0.5 7	place + places	3.4 7 × 1.4	places + place	1 6.5 × 2.8	place + place
		places		places		places
5.	4 5.4 ×4.0 2	place + places	1.5 4 × 1 0.6	places + place	4.6 8 ×3.1 2	places + places
		places		places		places

Multiply. Write zeros as needed.

	a	b	c	d
1.	0.3 ×0.6	0.0 3 × 0.6	9.8 ×0.5	6.3 ×0.0 4
2.	0.0 8 ×0.0 4	0.0 0 6 × 0.4	1.3 7 × 0.8	1.3 7 ×0.0 0 8
3.	0.0 1 5 × 0.1 4	0.0 7 5 × 0.2 2	1 5.5 ×2.1 2	7.0 5 ×2.0 4

Line up the digits. Then multiply. Write zeros as needed.

a	b	c
4. 0.43 × 0.02 = _____	0.206 × 0.37 = _____	8.79 × 6.08 = _____

0.43
×0.02

→ **MIXED PRACTICE**

Find each answer.

	a	b	c	d
1.	7.0 0 5 0.2 5 1 6. + 3.1	1.0 9 0.0 3 6 0.1 7 +2.4	1 7.6 4 9 3.0 5 4 0.0 1 6 + 0.1 9	2.0 5 3 0.1 7 6 0.0 3 4 +2.9 3 4
2.	1.9 4 3 −0.8 4 9	0.7 3 4 −0.2 7 5	1 9.0 6 − 9.9 7	2 0 5 −1 5 6.4
3.	$3 0 1.0 5 − 1 9 6.4 8	$6 2 0.0 4 − 1 4 8.9 9	$1 0 0 0.4 4 − 9 0 9.7 5	$8 7 0.0 0 − 6 7 0.9 9
4.	$9\frac{5}{8}$ $-\frac{1}{2}$	$16\frac{5}{6}$ $-\frac{1}{3}$	$71\frac{1}{4}$ $-\frac{1}{2}$	$7\frac{2}{7}$ $-\frac{2}{3}$

Dividing Decimals by Whole Numbers

To divide a decimal by a whole number, write the decimal point in the quotient directly above the decimal point in the dividend. Then divide as with whole numbers.

Find: 9.92 ÷ 16

Write a decimal point in the quotient.	Divide.
$16\overline{)9.92}$	$\begin{array}{r} 0.62 \\ 16\overline{)9.92} \\ \underline{9\,6}\downarrow \\ 32 \\ \underline{32} \\ 0 \end{array}$

Find: $48.96 ÷ 24

Write a decimal point in the quotient.	Divide.
$24\overline{)\$48.96}$	$\begin{array}{r} \$2.04 \\ 24\overline{)\$48.96} \\ \underline{48}\downarrow\downarrow \\ 96 \\ \underline{96} \\ 0 \end{array}$

PRACTICE

Divide.

	a	*b*	*c*	*d*
1.	$\begin{array}{r} 8.2 \\ 8\overline{)65.6} \\ \underline{64}\downarrow \\ 16 \\ \underline{16} \\ 0 \end{array}$	$5\overline{)\$3.45}$	$3\overline{)8.28}$	$7\overline{)0.784}$
2.	$\begin{array}{r} 0.04 \\ 61\overline{)2.44} \\ \underline{244} \\ 0 \end{array}$	$39\overline{)\$58.50}$	$46\overline{)9.338}$	$14\overline{)\$43.96}$
3.	$7\overline{)29.12}$	$4\overline{)\$16.48}$	$71\overline{)1.278}$	$22\overline{)0.154}$

Set up the problem. Then divide.

	a	*b*	*c*
4.	$22.5 ÷ 15 = $ _____	$\$6.03 ÷ 9 = $ _____	$114.8 ÷ 82 = $ _____

$15\overline{)22.5}$

Dividing Decimals by Decimals

To divide by a decimal, change the divisor to a whole number by moving the decimal point. Move the decimal point in the dividend the same number of places. Then divide.

Remember, write a decimal point in the quotient directly above the new decimal point position in the dividend.

Find: $4.34 ÷ 0.7

Move each decimal point 1 place.	Divide.
0.7)$4.34	6.2 7)43.4 42↓ 1 4 1 4 0

Find: 0.0713 ÷ 0.23

Move each decimal point 2 places.	Divide.
0.23)0.0713	0.31 23)007.13 69↓ 23 23 0

PRACTICE

Divide.

	a	*b*	*c*	*d*
1.	4.8 1.7)8.1 6 6 8↓ 1 3 6 1 3 6 0	0.5)2.6 4 5	4.6)0.0 1 3 8	3.9)$5 3.4 3
2.	29.9 0.1 6)4.7 8 4 3 2↓ 1 5 8 1 4 4↓ 1 4 4 1 4 4 0	0.2 4)1.4 8 8	0.0 8)$6.4 8	0.5 7)2.5 6 5

Set up the problem. Then divide.

	a	*b*	*c*
3.	1.854 ÷ 0.9 = _____	0.91 ÷ 1.3 = _____	$15.18 ÷ 0.33 = _____

0.9)1.8 5 4

Dividing Whole Numbers by Decimals

To divide a whole number by a decimal number, change the decimal number to a whole number. Move the decimal point in the dividend the same number of places. You will need to add one or more zeros. Then divide.

Find: 102 ÷ 1.7

Move each decimal point 1 place.	Divide.
$1.7\overline{)102.0}$	$17\overline{)1020}$ gives 60 $\underline{102}{\downarrow}$ 00

Find: $230 ÷ 0.25

Move each decimal point 2 places.	Divide.
$0.25\overline{)\$230.00}$	$025\overline{)\$23000}$ gives $\$920$ $\underline{225}{\downarrow}$ 50 $\underline{50}{\downarrow}$ 00

PRACTICE

Divide. Write zeros as needed.

	a	*b*	*c*	*d*
1.	$0.6\overline{)96.0}$ → 160 $\underline{6}{\downarrow}$ 36 $\underline{36}{\downarrow}$ 00	$2.4\overline{)72}$	$1.8\overline{)9}$	$0.9\overline{)\$306}$
2.	$0.34\overline{)51.00}$ → 150 $\underline{34}$ 170 $\underline{170}$ 00	$0.17\overline{)85}$	$0.08\overline{)\$52}$	$0.32\overline{)128}$
3.	$0.2\overline{)\$7}$	$0.5\overline{)13}$	$0.14\overline{)84}$	$1.6\overline{)\$224}$

Set up the problem. Then divide. Write zeros as needed.

	a	*b*	*c*
4.	$\$117 ÷ 7.8 =$ _____ $7.8\overline{)\$117}$	$162 ÷ 1.5 =$ _____	$328 ÷ 0.4 =$ _____

WORKING WITH DECIMALS

Decimal Quotients

Sometimes when you divide, the divisor will be larger than the dividend. To divide, add a decimal point and zeros as needed to the dividend. Continue to divide until the remainder is zero. In some cases, you may never have a remainder of zero. When dividing money, round the quotient to the nearest cent. Remember, zeros may be needed in the quotient also.

You can use this method to change some fractions to decimals by dividing the numerator by the denominator.

Find: $19 ÷ 300

Add a decimal point and zeros to the dividend.

$$300 \overline{)\ \$19.00}$$

Divide. Place a zero in the quotient.

$$
\begin{array}{r}
\$0.0633 \\
300 \overline{)\ \$19.0000} \\
18\ 00 \\
\hline
1\ 000 \\
900 \\
\hline
1000 \\
900 \\
\hline
100
\end{array}
$$

Round $0.0633 to $0.06.

Change $\frac{8}{125}$ to a decimal.

Divide the numerator by the denominator. Add a decimal point and zeros to the dividend.

$$125 \overline{)\ 8.00}$$

Divide until the remainder is zero.

$$
\begin{array}{r}
0.064 \\
125 \overline{)\ 8.000} \\
7\ 50 \\
\hline
500 \\
500 \\
\hline
0
\end{array}
$$

$\frac{8}{125} = 0.064$

PRACTICE

Divide. Write zeros as needed.

	a	b	c	d
1.	$40 \overline{)20.0}$ quotient 0.5, 200, 0	$100 \overline{)\$30}$	$5 \overline{)2}$	$20 \overline{)\$15}$
2.	$50 \overline{)1}$	$500 \overline{)25}$	$5 \overline{)\$4}$	$20 \overline{)19}$

Use division to change each fraction to a decimal.

	a	b	c	d
3.	$\frac{1}{4}$ $4 \overline{)1.00}$ quotient 0.25, 8, 20, 20, 0	$\frac{1}{8}$	$\frac{1}{40}$	$\frac{3}{5}$
4.	$\frac{6}{12}$	$\frac{6}{8}$	$\frac{2}{500}$	$\frac{15}{60}$

129

WORKING WITH DECIMALS
Multiplying and Dividing by Powers of 10

To multiply decimals by powers of ten, move the decimal point in the product to the right as many places as there are zeros in the multiplier.

Remember, sometimes you might need to write zeros in the product in order to move the decimal point the correct number of places.

Study these examples.

$10 \times 0.89 = 8.9$ $100 \times 0.73 = 73$ $1000 \times 0.52 = 520$
$10 \times 8.9 = 89$ $100 \times 7.3 = 730$ $1000 \times 5.2 = 5200$

To divide a decimal by a power of ten, move the decimal point in the dividend to the left as many places as there are zeros in the divisor.

Remember, sometimes you might need to write zeros in the quotient in order to correctly insert the decimal point.

Study these examples.

$0.89 \div 10 = 0.089$ $0.73 \div 100 = 0.0073$ $0.52 \div 1000 = 0.00052$
$8.9 \div 10 = 0.89$ $7.3 \div 100 = 0.073$ $5.2 \div 1000 = 0.0052$

PRACTICE

Multiply or divide. Write zeros as needed.

	a	b	c
1.	$7.5 \times 10 =$ _____75_____	$46 \times 10 =$ _____	$0.07 \times 10 =$ _____
2.	$100 \times 0.7 =$ _____	$100 \times 4.6 =$ _____	$0.075 \times 100 =$ _____
3.	$0.5 \div 10 =$ _____0.05_____	$8 \div 1000 =$ _____	$1.25 \div 100 =$ _____
4.	$12.5 \div 100 =$ _____	$0.125 \div 1000 =$ _____	$14.92 \div 100 =$ _____
5.	$6.2 \times 1000 =$ _____	$642.15 \div 10 =$ _____	$642.15 \div 100 =$ _____
6.	$3.15 \times 1000 =$ _____	$0.048 \times 100 =$ _____	$0.048 \div 100 =$ _____
7.	$0.375 \div 10 =$ _____	$3.75 \div 10 =$ _____	$37.5 \div 10 =$ _____
8.	$375 \div 10 =$ _____	$0.375 \times 1000 =$ _____	$0.007 \times 1000 =$ _____
9.	$719.35 \times 100 =$ _____	$16.147 \times 1000 =$ _____	$14.92 \div 1000 =$ _____
10.	$267.18 \div 100 =$ _____	$2.6718 \div 1000 =$ _____	$2.6718 \times 1000 =$ _____

WORKING WITH DECIMALS

Estimating Products and Quotients

To estimate decimal products or quotients, round each number to the same place. Then multiply or divide the rounded numbers.

Estimate: 15.3 × 2.8

Round each number.	Multiply.
15.3 → *15*	*15*
2.8 → *3*	*× 3*
	45

Estimate: 360.41 ÷ 2.89

Round each number.	Divide.
360.41 → *360*	*120*
2.89 → *3*	*3) 360*
	3↓
	06
	6↓
	00
	00
	0

PRACTICE

Estimate each product.

	a	b	c
1.	$\begin{array}{r} 31.75 \to 3\,2 \\ \times\ \ \ 2.2 \to \times\ \ 2 \\ \hline 6\,4 \end{array}$	$\begin{array}{r} 15.6 \to \\ \times\ \ 3.5 \to \\ \hline \end{array}$	$\begin{array}{r} 23.8 \to \\ \times\ \ 4.7 \to \\ \hline \end{array}$
2.	$\begin{array}{r} 9.5 \to \\ \times 13.42 \to \\ \hline \end{array}$	$\begin{array}{r} 1.17 \to \\ \times\ \ 8.1 \to \\ \hline \end{array}$	$\begin{array}{r} 12.69 \to \\ \times\ \ \ 6.2 \to \\ \hline \end{array}$

Estimate each quotient.

	a	b	c
3.	$4.9\overline{)19.95} \to 5\overline{)20}$ ⁴	$6.2\overline{)12.17} \to$	$9.95\overline{)109.5} \to$
4.	$22.15\overline{)65.79} \to$	$3.6\overline{)23.8} \to$	$2.98\overline{)74.97} \to$

Estimate each answer.

	a	b	c
5.	45.8×9.3	$30.27 \div 10.14$	31.996×4.2
	$\begin{array}{r} 46 \\ \times\ 9 \\ \hline \end{array}$		

WORKING WITH DECIMALS

Practice Multiplying and Dividing Decimals

Estimate the answer. Then multiply or divide. Use your estimate to check your answer.

Multiply.

	a	b	c	d
1.	0.3 × 8	43 × 0.2 4	0.3 5 1 × 8 6	0.6 7 3 9 × 7
2.	7 5 × 0.4 8	0.0 3 × 2	1 4 × 0.0 0 7	0.0 0 2 × 4
3.	3.1 4 × 1 8	0.6 × 0.3	0.3 × 0.2	0.5 8 × 0.6

Divide.

4.

$$4\overline{)9.2} \qquad 0.3\overline{)2\,4\,7.8} \qquad 0.0\,2\overline{)5\,2\,1.5\,6} \qquad 0.0\,0\,6\overline{)7\,4.8\,9\,8}$$

5.

$$7\overline{)8.9\,6} \qquad 0.5\overline{)9.2\,5} \qquad 0.7\,9\overline{)4.6\,6\,1} \qquad 0.0\,1\,8\overline{)0.4\,5\,5\,4}$$

6.

$$2\overline{)5.3\,2\,8} \qquad 0.8\overline{)0.8\,9\,6} \qquad 0.5\,6\overline{)5.0\,1\,7\,6} \qquad 0.0\,0\,7\overline{)6.5\,3\,9\,1\,2}$$

More Practice Multiplying and Dividing Decimals

After you find an answer, check with a calculator.

PRACTICE
Multiply.

	a	b	c	d
1.	15.3	28.7	12.615	8.7
	$\times\ \ 8.1$	$\times\ 0.24$	$\times\ \ \ \ \ 25$	$\times\ 0.48$

2.	0.21	0.56	0.05	16.2
	$\times\ \ \ 0.4$	$\times\ 0.37$	$\times\ 0.01$	$\times\ 0.045$

3.	34.89	0.147	3.1416	0.059
	$\times\ 0.875$	$\times\ \ \ \ 0.03$	$\times\ \ \ \ 0.75$	$\times\ 0.064$

Divide.

4.

$8\,\overline{)\,0.736}$ $\qquad$ $0.2\,\overline{)\,0.0034}$ $\qquad$ $0.03\,\overline{)\,0.0009}$ $\qquad$ $0.231\,\overline{)\,0.00924}$

5.

$36\,\overline{)\,91.44}$ $\qquad$ $1.2\,\overline{)\,108.72}$ $\qquad$ $1.44\,\overline{)\,135.072}$ $\qquad$ $0.048\,\overline{)\,60}$

6.

$8\,\overline{)\,5.000}$ $\qquad$ $0.6\,\overline{)\,12.0}$ $\qquad$ $0.25\,\overline{)\,50}$ $\qquad$ $0.125\,\overline{)\,53.75}$

PROBLEM-SOLVING STRATEGY
Identify Extra Information

Some problems may include more facts than you need to solve the problem. Often you must read the problem several times to decide which facts are needed and which facts are extra. Crossing out extra facts can help you solve the problem. Use the remaining facts to solve the problem.

Read the problem.

Steve is 26 years old. He has worked as an electrician for $4\frac{1}{2}$ years.

When he works overtime, he earns 1.5 times his regular hourly wage. He earns $14.40 per hour working from 9 AM to 4:30 PM. What is his hourly wage when he works overtime?

Decide which facts are needed.

Steve earns $14.40 per hour.

Steve earns 1.5 times his hourly wage when he works overtime.

Decide which facts are extra.

Steve is 26 years old.

He has worked for $4\frac{1}{2}$ years.

He works from 9 AM to 4:30 PM.

Solve the problem.

$$
\begin{array}{r}
\$1\,4.4\,0 \\
\times \quad 1.5 \\
\hline
7\,2\,0\,0 \\
1\,4\,4\,0 \\
\hline
2\,1.6\,0\,0
\end{array}
$$

Steve's overtime wage is $21.60 per hour.

In each problem, cross out the extra facts. Then solve.

1. Rosa bought 3 pens at 89¢ each, 2 videotapes at $4.49, and 2 records at $7.98. How much did she spend on videotapes?

$$
\begin{array}{r}
\$\,4.4\,9 \\
\times \quad 2
\end{array}
$$

2. Leroy is 22 years old. His car averages 31 miles per gallon. His car payments are $165.32 per month, and he has 36 more payments to make. How old will he be when he pays off his car?

Answer _____

Answer _____

In each problem, cross out the extra facts. Then solve.

3. Olivia drove her 1988 sedan for 2.25 hours at an average speed of 52 miles per hour. She used 5 gallons of gas priced at $1.12 per gallon. How far did she drive?

Answer _____

4. Renaldo takes the bus 2.6 miles to work each day. The trip takes 35 minutes. He earns $24,480 per year. What is Renaldo's monthly salary?

Answer _____

5. Ray bought 160 feet of 12-gauge wire for $20.80. The wire came in 4 rolls, each one 40 feet in length. What was the price per foot of the wire?

Answer _____

6. Lucy needed 256 feet of fencing to enclose her yard. The fencing came in 4 styles. The picket fencing would not be available for 6–8 weeks. How much will her fence cost at $1.19 per foot?

Answer _____

7. Tickets for the 3-hour concert cost $15. There were 18,207 people at the concert. The 5 band members played 20 songs. How much money was collected for the sale of tickets?

Answer _____

8. Elizabeth was planning a trip to Colorado. The plane trip would take 2 hours and 47 minutes and cost $258. Her hotel costs $54 per night. What will her hotel bill be for 6 nights?

Answer _____

9. Ryan went grocery shopping. He was at the store for 45 minutes. He spent $36.52. He bought enough ingredients to make 3 quarts of spaghetti sauce. How much change did he receive from $40.00?

Answer _____

10. Deena needed two new computers for her office. A box of disks cost $89. The salesperson at the computer store said that the computers would be delivered in 5 working days. How much will 4 boxes of disks cost?

Answer _____

Applications

Solve.

1. Diane worked 38.6 hours last week. At $5.55 per hour, how much did she earn?

   Answer _____

2. Yesterday we took a trip by car and averaged 75.4 kilometers an hour. We drove for 8.5 hours. How far did we drive?

   Answer _____

3. Karen bought 3 cassette tapes for $9.88 each. What was the total?

   Answer _____

4. Nick bought 5 blades for his circular saw at $12.67 each. How much did he pay for the 5 blades?

   Answer _____

5. Mr. Washington bought 43.5 meters of pipe for $4.28 per meter. How much was the total?

   Answer _____

6. C. W. Dillard is a carpenter and receives $15.25 an hour. She works an average of 40 hours a week. What are her weekly earnings?

   Answer _____

7. Ms. Lennox is paid $5.10 an hour, and 1.5 times that rate for every hour worked beyond 40. How much will she be paid for working 42 hours?

   Answer _____

8. Dennis is making monthly car payments of $185.37. How much does he pay on the car in a year?

   Answer _____

Unit 5 Review

Write as decimals.

	a	b	c	d
1.	$\frac{1}{10} =$ _____	$\frac{3}{100} =$ _____	$\frac{1}{5} =$ _____	$\frac{3}{4} =$ _____
2.	$\frac{1}{8} =$ _____	$\frac{5}{8} =$ _____	$1\frac{1}{4} =$ _____	$2\frac{3}{5} =$ _____

Write as a fraction or a mixed number.

	a	b	c	d
3.	$1.1 =$ _____	$1.5 =$ _____	$0.75 =$ _____	$5.50 =$ _____

Find each answer.

	a	b	c	d
4.	7.7 6 6.6 7 +4.3 9	1.9 2 3 2.7 4 9 +1.6 3 7	0.0 9 0.5 4 +0.9 7 8	0.0 1 8 0.2 0 9 +40.0 9
5.	7.9 4 −4.5 6	0.5 0 6 −0.1 8 9 2	3 0 6.0 9 − 4 6.4 5	7.4 7 6 3 −6.4 7 6 7
6.	1.2 5 ×0.0 4	3.0 8 × 1 2	3.5 ×2.4	2.0 7 5 × 0.0 8
7.	6)0.3 6	1 2)1.4 4	0.1 2)1 4 4	0.6)3.6

Multiply or divide.

	a	b	c
8.	$2.25 \times 10 =$ _____	$0.225 \times 1000 =$ _____	$22.5 \times 100 =$ _____
9.	$35.2 \div 10 =$ _____	$18.6 \div 100 =$ _____	$149.2 \times 100 =$ _____

PERCENTS

Meaning of Percent

The symbol % is read as *percent*. Percent means *per hundred* or *out of one hundred*. Therefore, 35% means 35 per hundred or 35 out of 100, or 35 hundredths.

By writing the percent as hundredths, we can write a percent as a fraction or as a decimal.

To change a *percent to a decimal*, move the decimal point 2 places to the left and drop the percent sign (%). Write zeros as needed.

EXAMPLES

$$35\% = 0.35 \qquad 6\% = 0.06 \qquad 480\% = 4.8$$

To change a *percent to a fraction*, place the percent over 100 and drop the % sign. Simplify.

EXAMPLES

$$35\% = \frac{35}{100} = \frac{7}{20} \qquad 6\% = \frac{6}{100} = \frac{3}{50} \qquad 480\% = \frac{480}{100} = 4\frac{4}{5}$$

PRACTICE

Change each percent to a decimal and then to a fraction. Simplify.

a   *b*

1. $82\% = $ $0.82 = \frac{82}{100} = \frac{41}{50}$ $7\% = $

2. $1\% = $ $142\% = $

3. $95\% = $ $55\% = $

4. $109\% = $ $73\% = $

5. $98\% = $ $12\% = $

6. $4\% = $ $44\% = $

7. $175\% = $ $83\% = $

8. $26\% = $ $137\% = $

PERCENTS

Changing Decimals and Fractions to Percents

To change a *decimal to a percent*, move the decimal point 2 places to the right and write a percent symbol. Write zeros as needed.

EXAMPLES

$$0.825 = 82.5\% \qquad 0.03 = 3\% \qquad 0.4 = 40\%$$

PRACTICE

Change the following to percents.

	a	*b*	*c*
1.	$0.30 =$ ___30%___	$0.08 =$ _____	$0.45 =$ _____
2.	$0.91 =$ _____	$0.56 =$ _____	$1.49 =$ _____
3.	$0.73 =$ _____	$0.672 =$ _____	$0.02 =$ _____
4.	$3.25 =$ _____	$0.09 =$ _____	$0.7 =$ _____
5.	$1.333 =$ _____	$0.54 =$ _____	$0.62 =$ _____

To change a *fraction to a percent*, first change the fraction to a decimal by dividing the numerator by the denominator. Then rewrite the decimal quotient as a percent.

Write $\frac{3}{4}$ as a percent.

Divide.

$$\begin{array}{r} 0.75 = 75\% \\ 4\overline{)3.00} \end{array}$$

Write $\frac{3}{20}$ as a percent.

Divide.

$$\begin{array}{r} 0.15 = 15\% \\ 20\overline{)3.00} \end{array}$$

PRACTICE

Change the following to percents.

	a	*b*	*c*
6.	$\frac{3}{8} =$ ___37.5%___	$\frac{2}{5} =$ _____	$\frac{7}{10} =$ _____
7.	$\frac{21}{100} =$ _____	$\frac{5}{16} =$ _____	$\frac{1}{4} =$ _____
8.	$\frac{3}{5} =$ _____	$\frac{7}{20} =$ _____	$\frac{5}{12} =$ _____

PERCENTS

Interchanging Fractions, Decimals, and Percents

PRACTICE

Fill in the blanks below as illustrated in the first example.

1. $\frac{6}{10}$ = _0.6_ = _60%_ 　　-Work Space-

2. $\frac{1}{4}$ = _0.25_ = _____

3. _____ = _____ = _20%_

4. $\frac{17}{20}$ = _____ = _____

5. _____ = _0.5_ = _____

6. _____ = _____ = _75%_

Solve.

7. The Crowe family spends $27 out of every $100 earned on rent. What percent of their income does the Crowe family spend on rent?

Answer _____

8. The sales tax in one city is $7\frac{1}{2}\%$. Write $7\frac{1}{2}\%$ as a decimal.

Answer _____

9. A suit is on sale for 10% off the regular price. Write the decimal the salesperson would use to figure the amount of the discount.

Answer _____

10. At Souper Subs, 8 out of the 25 employees work part time. What percent of the employees work part time?

Answer _____

11. Juanita saves 15% of her salary. What fraction of her income does Juanita save?

Answer _____

12. According to a recent survey, five out of eight young people do some kind of exercise each week. Write this fraction as a decimal.

Answer _____

PERCENTS

Percents Greater Than 100%

Large percents are used just like the percents we have been using. To change a percent to a decimal, move the decimal point 2 places to the left. Drop the percent sign.

EXAMPLES

$$200\% = 2.00 = 2 \qquad 375\% = 3.75 \qquad 105\frac{3}{10}\% = 105.3\% = 1.053$$

PRACTICE

Change each percent to a decimal.

a	b	c
1. 205% = _2.05_	500% = _____	175% = _____
2. 999.9% = _____	320% = _____	$101\frac{3}{10}\%$ = _____
3. 450% = _____	609% = _____	1100% = _____
4. 807% = _____	$725\frac{1}{2}\%$ = _____	198% = _____

Solve.

5. A clothing store has a 210% markup on the price of their clothing. Write the decimal used to figure the price of their clothing.

Answer _____

6. Larry's test scores have improved 152%. Write this percent as a decimal.

Answer _____

7. The profits for Fun Times, Inc. have increased 333% since the company started. Write this amount as a decimal.

Answer _____

8. Doreen's savings have increased by 135%. What number would you use to figure the amount of money in Doreen's account?

Answer _____

Percents Less Than 1%

Small percents can be used just like the large percents we have been using. Proceed as you did with large percents. To solve problems using percents smaller than 1%, change the fraction to a decimal.

EXAMPLES

$$1\% = \frac{1}{100} = 100 \overline{)1.00}^{\,0.01} \qquad \frac{3}{4}\% = 0.75\% = \frac{0.75}{100} = 100 \overline{)0.7500}^{\,0.0075}$$

PRACTICE

Change each percent to a decimal.

	a	*b*	*c*
1.	0.5% = ___0.005___	0.33% = _____	0.25% = _____
2.	0.4% = _____	0.41% = _____	0.05% = _____
3.	$\frac{1}{4}$% = _____	$\frac{1}{2}$% = _____	$\frac{2}{5}$% = _____
4.	$\frac{4}{5}$% = _____	$\frac{3}{8}$% = _____	$\frac{7}{8}$% = _____

Solve.

5. Laurel's weight decreased by 0.9% since she last weighed herself. Write the decimal you would use to find Laurel's new weight.

Answer _____

6. The measurements were off by 0.64%. Write the number you would use to show the error.

Answer _____

7. Joanne's test scores improved by 0.55%. What number would you use to find Joanne's test scores?

Answer _____

8. Which is greater, 0.45% or $\frac{5}{12}$? Write each as a decimal and compare.

Answer _____

PROBLEM SOLVING

Applications

Solve.

1. The average household spends $0.35 of every dollar earned on housing costs. What percent is this amount?

   Answer _____

2. Three out of four students passed the exam. What percent of the students passed the exam?

   Answer _____

3. Real estate taxes in the city increased by 7.25%. What decimal number would you use to figure the new real estate taxes?

   Answer _____

4. In the last election, 36% of the people voted "Yes" on Proposition 6. What fraction of the people is this?

   Answer _____

5. The bank used 0.0975 to figure the interest charge on the McCall's mortgage. What percent is this?

   Answer _____

6. On the average, 17 out of every 100 viewing minutes is taken up by commercials. What is this average as a percent?

   Answer _____

7. Sales tax in the city is $8\frac{1}{2}\%$. Write the number used to figure the sales tax on a purchase.

   Answer _____

8. Home improvement costs have increased by $\frac{2}{5}$ in the last few years. By what percent have home improvement costs increased?

   Answer _____

PROBLEM-SOLVING STRATEGY

Use a Bar Graph

A graph is a picture of data (factual information). A bar graph shows information in the shape of bars. Use a bar graph to answer questions about the data.

This bar graph shows the number of cups of each flavor of frozen yogurt sold at The Freezer during the summer.

Read the problem.
How many more cups of strawberry were sold than cups of chocolate?

Read the graph for data.
Use these steps when finding the needed data.

1. Locate the flavor on the horizontal scale along the bottom of the graph.

2. For each flavor, move to the top of the bar and back to the vertical scale along the side of the graph.

3. Write down the data.

   Data: 300 cups of strawberry were sold.
   175 cups of chocolate were sold.

Solve the problem.

$$300 - 175 = 125$$

There were 125 more cups of strawberry sold than of chocolate.

Solve.

1. How many cups of yogurt were sold in all? Answer _____	**2.** Which flavors of yogurt had sales of less than two hundred? Answer _____
3. Which flavor of yogurt had the least number of sales? Answer _____	**4.** How many more cups of vanilla were sold than cups of peach? Answer _____
5. How many fewer cups of chocolate were sold than cups of vanilla? Answer _____	**6.** How many more cups of strawberry were sold than cups of peach? Answer _____

144

Solve. Use the bar graph below.

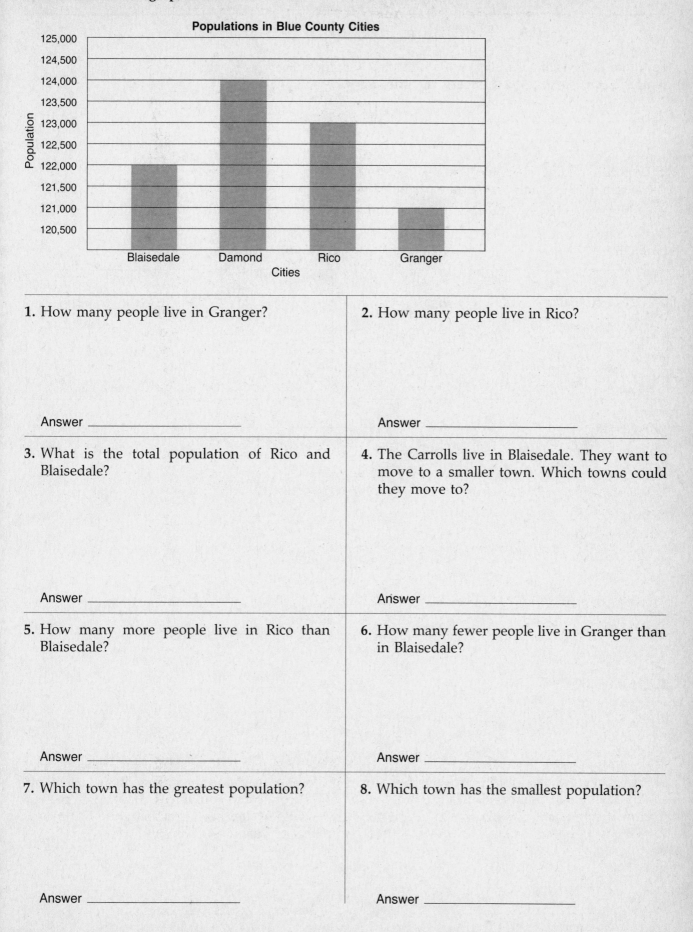

Populations in Blue County Cities

1. How many people live in Granger?

   Answer _____

2. How many people live in Rico?

   Answer _____

3. What is the total population of Rico and Blaisedale?

   Answer _____

4. The Carrolls live in Blaisedale. They want to move to a smaller town. Which towns could they move to?

   Answer _____

5. How many more people live in Rico than Blaisedale?

   Answer _____

6. How many fewer people live in Granger than in Blaisedale?

   Answer _____

7. Which town has the greatest population?

   Answer _____

8. Which town has the smallest population?

   Answer _____

Finding a Percent of a Number

To find a percent of a number, write a percent sentence. Every percent sentence consists of three numbers: the rate, the whole, and the part.

$$20\% \text{ of } 48 = 9.6$$

rate whole part

If the part is missing in a percent problem, solve by first changing the *rate* to a decimal. Then multiply the rate by the *whole*.

Remember, "of" means multiply.

Find: 25% of 84

$$25\% \times 84 = ?$$
$$0.25 \times 84 = 21$$

$$
\begin{array}{r}
84 \\
\times\ 0.25 \\
\hline
420 \\
168 \\
\hline
21.00
\end{array}
$$

Find: 105% of 280

$$105\% \times 280 = ?$$
$$1.05 \times 280 = 294$$

$$
\begin{array}{r}
280 \\
\times\ 1.05 \\
\hline
1400 \\
000 \\
280 \\
\hline
294.00
\end{array}
$$

PRACTICE

Change each percent to a decimal. Solve.

	a	b
1.	50% of 90	300% of 60
	$0.5 \times 90 = 45$	
2.	80% of 120	75% of 80
3.	40% of 75	90% of 200
4.	250% of 100	99% of 55

Solve.

5. Rachel decided to save 15% of the money she earned. In one month, Rachel earned $84. How much money did she save?

6. Walter read an advertisement for a 30%-off sale on stereo equipment. How much money would Walter save if he bought a CD player that normally sold for $219?

Answer _____

Answer _____

PERCENTS

Finding a Percent of a Number

Another way to find the missing *part* in a percent problem is to use fractions rather than decimals. Change the *rate* and the *whole* to fractions. Multiply the fractions.

Find: 20% of 50

$$20\% \times 50 = ?$$

$$\frac{1}{\cancel{5}} \times \frac{\cancel{50}^{\,10}}{1} = \frac{10}{1} = 10$$

Find: 75% of $1\frac{3}{4}$

$$75\% \times 1\frac{3}{4} = ?$$

$$\frac{3}{4} \times \frac{7}{4} = \frac{21}{16} = 1\frac{5}{16}$$

PRACTICE

Change each percent to a fraction. Solve.

a	*b*

1. 25% of 16

$$\frac{1}{\cancel{4}} \times \frac{\cancel{16}^{\,4}}{1} = \frac{4}{1} = 4$$

50% of $1\frac{1}{3}$

2. 85% of 40

40% of $3\frac{1}{3}$

3. 60% of 55

80% of $3\frac{3}{4}$

4. 6% of 20

70% of $6\frac{2}{3}$

Solve.

---

5. A sweater that sells for $66 was on sale for $33\frac{1}{3}$% off. How much would you save on the sweater? (Hint: $33\frac{1}{3}\% = \frac{1}{3}$.)

6. A survey of college students showed that $66\frac{2}{3}$% of the 600 students surveyed studied a foreign language. How many students studied a foreign language? (Hint: $66\frac{2}{3}\% = \frac{2}{3}$.)

Answer _____

Answer _____

PERCENTS

Finding What Percent One Number Is of Another

To find the *rate* in a percent problem, write a percent sentence. Divide the *part* by the *whole*. Then write the decimal answer as a percent.

What percent of 75 is 60?

$$? \% \times 75 = 60$$
$$? = 60 \div 75$$
$$0.8 = 60 \div 75$$
$$0.8 = 80\%$$

What percent of 160 is 4?

$$? \% \times 160 = 4$$
$$? = 4 \div 160$$
$$0.025 = 4 \div 160$$
$$0.025 = 2.5\%$$

PRACTICE

Find the rate.

         a   *b*

1. What percent of 8 is 40?

$$? = 40 \div 8$$
$$5 = 40 \div 8$$
$$5 = 500\%$$

What percent of 25 is 15?

2. What percent of 48 is 7.2?

What percent of 23 is 6.9?

3. What percent of 180 is 81?

What percent of $35 is $5.25?

4. What percent of 400 is 268?

What percent of 112 is 140?

Solve.

5. Jerry made a down payment of $3850 on a car that cost $11,000. What percent of the price of the car was his down payment?

Answer _____

6. Ellen got a $10 raise this week. Her salary before the raise was $250 a week. By what percent did her salary increase?

Answer _____

7. The Schultes save $30 a week for their vacation. They have a combined income of $400 a week. What percent of their earnings do they save for vacation?

Answer _____

8. There were 480 students at Kelley School last year. This year there are only 408. By what percent did the enrollment decrease?

Answer _____

PERCENTS

Finding a Number When a Percent of It Is Known

To find the *whole* in a percent problem, write a percent sentence. Change the *rate* to a decimal. Divide the *part* by the decimal.

12% of what number is 18?

$$12\% \times ? = 18$$
$$? = 18 \div 0.12$$
$$150 = 18 \div 0.12$$

30% of what number is 180?

$$30\% \times ? = 180$$
$$? = 180 \div 0.30$$
$$600 = 180 \div 0.30$$

PRACTICE

Change each percent to a decimal. Solve.

a	*b*

1. 25% of what number is 17?

$$? = 17 \div 0.25$$
$$68 = 17 \div 0.25$$

32% of what number is 40?

2. 80% of what number is 64?

135% of what number is 270?

3. 60% of what number is 33?

45% of what number is 90?

4. 10% of what number is 73?

75% of what number is 120?

Solve.

---

5. Peter bought a sweater for $32. This was 80% of the original price. What was the original price?

Answer _____

6. Katie got a 5% discount for paying cash on her furniture purchase. She paid $2850 for the furniture. What would the price have been if she had charged her purchase?

Answer _____

---

7. The distance by boat from New York City to San Francisco is 5200 miles by way of the Panama Canal. This is 40% of the distance by way of the Strait of Magellan. How far is it by way of the Strait of Magellan?

Answer _____

8. Adelena got a score of 90% on a true-false test. She answered 36 questions correctly. How many questions were on the test?

Answer _____

PERCENTS

Simple Interest

Interest is a charge that is paid for borrowed money. To find the simple interest on a loan, multiply the amount borrowed, or principal, times the annual *rate* of interest (a percent) times the number of years the loan is for. Change part of a year to a decimal.

EXAMPLE: Find the simple interest on a loan of $2000 for 3 years at a rate of 6% per year.

$$\text{Interest} = \text{Principal} \times \text{Rate} \times \text{Time}$$
$$I = prt$$
$$I = \$2000 \times 0.06 \times 3$$
$$I = \$360$$

GUIDED PRACTICE

Find the simple interest.

a	*b*

1. $500 at 5.5% for $\frac{1}{2}$ year

$I = \$500 \times 0.055 \times 0.5$

$I = 500 \times 0.0275$
$I = \$13.75$

$225 at $6\frac{1}{2}$% for 1 year

$I = \$225 \times 0.065 \times 1$

2. $800 at 18% for 1 year
$I = \$800 \times 0.18 \times 1$

$1000 at 5% for 2 years
$I = \$1000 \times 0.05 \times 2$

3. $400 at 8.25% for 3 months
$I = \$400 \times 0.0825 \times 0.25$

$150 at 12% for 2 years
$I = \$150 \times 0.12 \times 2$

4. $2000 at 7% for 5 years
$I = \$2000 \times 0.07 \times 5$

$850 at 4% for $\frac{1}{2}$ year
$I = \$850 \times 0.04 \times 0.5$

Find the simple interest. Round to the nearest cent as needed.

a	*b*
1. $3460 at 6.5% for 6 months (Hint: Change 6 months to 0.5 year.)	$75 at 5.25% for 1 year
2. $225 at 4.75% for $1\frac{1}{4}$ years (Hint: Change $1\frac{1}{4}$ years to 1.25 year.)	$3210 at 6% for 3 years
3. $615 at 6.5% for 2 years	$4000 at $6\frac{7}{10}$% for 4 years
4. $3750 at $6\frac{1}{4}$% for 3 months	$525 at 5% for 15 months
5. $2940 at 4.8% for 2.25 years	$465 at 4.75% for $1\frac{1}{2}$ years
6. $3500 at 5.75% for 6 years	$11,500 at 8.25% for 3 years

MIXED PRACTICE

Find each answer.

a	*b*	*c*	*d*
1. $\begin{array}{r} 9\,1.3\,4 \\ +5\,2.9 \\ \hline \end{array}$	$\begin{array}{r} \$1.4\,2 \\ \times\ \ 1\,8\,0 \\ \hline \end{array}$	$\begin{array}{r} \$2\,7.0\,0 \\ -\ \ \ \ 1.9\,9 \\ \hline \end{array}$	$3\overline{)1.8\,9}$
2. $\begin{array}{r} 6\,3\,7 \\ \times1\,2\,8 \\ \hline \end{array}$	$\begin{array}{r} 7\,5\,4 \\ -6\,9\,9 \\ \hline \end{array}$	$\begin{array}{r} 2,8\,1\,3 \\ +4\,8,9\,7\,5 \\ \hline \end{array}$	$2\,3\overline{)6\,7\,8\,5}$

Percent of Increase

To find the current value or amount, first multiply the original amount by the percent of increase. Then add.

EXAMPLE: The rent for a 1-bedroom apartment was $275 per month. This year the rent went up 4%. How much is the apartment renting for this year?

original amount	$ 275	original amount	$275
percent of increase	× 0.04	amount of increase	+ 11
amount of increase	$ 11	current rent	$286

The apartment will rent for $286 this year.

Solve.

---

1. Alvin weighed 120 pounds 6 months ago. Since then his weight has increased by 2%. How much does Alvin weigh now?

$$
\begin{array}{r}
120 \\
\times\ 0.02 \\
\hline
2.40
\end{array}
\qquad
\begin{array}{r}
120 \\
+\quad 2.4 \\
\hline
122.4
\end{array}
$$

Answer _____ *122.4 pounds* _____

2. A local movie theater increased the price of admission by 20%. Tickets had sold for $5.25. What is the current ticket price?

Answer _____

---

3. Louise bought her condominium for $56,900. After 2 years, its value has increased 17%. What is the current value of Louise's condominium?

Answer _____

4. Junior college enrollment was up by 4% from last year. The previous year's enrollment was 9050. How many students were attending the junior college this year?

Answer _____

---

5. The price of an economy car went up 7% from last year. The car sold for $11,500 last year. What is the list price for the same car this year?

Answer _____

6. The Steak House increased their menu prices by 6%. A complete dinner had been $12.50. What was the new price for the dinner?

Answer _____

PERCENTS

Percent of Decrease

To find the current value or amount, first multiply the original amount by the percent of decrease. Then subtract.

EXAMPLE: Last year's sales figure at Klaus Clothiers was $850,000. This year the total sales figure decreased by 3%. What was this year's sales figure?

original figure	$850,000	original figure	$850,000
percent of decrease	× 0.03	amount of decrease	− 25,500
amount of decrease	$25,500.00	new sales figure	$824,500

This year's sales figure at Klaus Clothiers was $824,500.

Solve.

---

1. A school had 440 students last year. This year the enrollment decreased by 5%. How many students are attending the school this year?

$$440 \times 0.05 = 22.00$$
$$440 - 22 = 418$$

Answer _____ *418 students* _____

2. A store decreased the price of its best shirts by 20%. If the shirts normally sold for $20, what was the reduced price?

Answer _____

---

3. The value of a car that cost $11,000 decreased by 20% the first year. What was the value of the car after the first year?

Answer _____

4. The average number of points scored by the Ridgeview basketball team decreased by 8%. They had been averaging 75 points per game. What was their new average?

Answer _____

---

5. A city had a population of 420,000 in 1992. In 1993, its population had decreased by 2%. How many people were living in the city in 1993?

Answer _____

6. The Rinalli's average gas bill decreased by 4.5% when they moved. They had been spending about $43 per month for gas. How much are they now spending for gas? Round to the nearest cent.

Answer _____

PROBLEM-SOLVING STRATEGY

Use Logic

Some problems can be solved using logic. Using logic is like using clues to solve a mystery. First, read the problem and look for clues. Then, make a list or chart, or draw pictures to help you keep track of the clues. Next, solve the problem.

EXAMPLE 1

Read the problem.

The last names of Al, Beth, and Charlie are Alvarez, Jones, and Smith. Beth's last name is not Smith. Charlie's last name is Alvarez. What are the last names of the three people?

Look for clues.

Clue 1.   Beth's name is not Smith.

Clue 2.   Charlie's name is Alvarez.

Make a chart.

Fill in the chart using the clues you have found.

	Alvarez	Jones	Smith
Al	no	no	**YES**
Beth	no	**YES**	no
Charlie	yes	no	no

Solve the problem.

Al's last name is Smith.
Beth's last name is Jones.
Charlie's last name is Alvarez.

EXAMPLE 2

Read the problem.

Su Ling and Vicki live in separate even-numbered apartments. Each of them has a two-digit apartment number. The sum of the two digits of each apartment is 14. What apartments do the two women live in?

Look for clues.

Clue 1.   The women live in even-numbered apartments.

Clue 2.   The sum of the digits of their apartment numbers is 14.

Make a list.

The two digit numbers that have a sum of 14 are:
59, 68, 77, 86, and 95.
Find the basic addition facts that equal 14.

Solve the problem.

The two even-numbered apartments with a sum of 14 are 68 and 86.

Use logic to solve each problem.

1. A drawer contains 10 blue socks and 10 red socks. How many socks must you remove, without looking, to be sure you have two of the same color?

Answer _____

2. A dog, a cat, a fish, and a goat are named Spike, Mo, Buck, and Fab. Mo is not a dog or cat. Neither Spike nor Fab is a dog. The cat tries to catch Fab. Mo can't swim. What is each animal's name?

Dog _____

Cat _____

Fish _____

Goat _____

3. Tom's age is divisible by 15. His father is 65. His son is 22. How old is Tom?

Answer _____

4. An even number has 2 digits. The sum of the digits is 11. The difference between the digits is 3. What is the number?

Answer _____

PROBLEM SOLVING

Applications

Solve.

1. The selling price on athletic shoes is a 65% increase over cost. Run-Fast running shoes cost the store $18. What is the selling price of the shoes?

Answer _____

2. A customer saved $72 by buying a dishwasher on sale for 30% off the regular price. What was the regular price?

Answer _____

3. A videocassette recorder that normally sells for $339 is on sale for 25% off. How much would you save if you bought the recorder while it was on sale?

Answer _____

4. Ed's TV and Appliance requires a 20% down payment to hold a purchase. If the deposit was $172, what was the total cost of the item?

Answer _____

5. Several years ago, a desk sold for $124.25. Now the same desk sells for $175. What percent of increase is the new price?

Answer _____

6. The sales tax on a television was $13.93. The sales tax rate is 7%. What is the selling price of the television?

Answer _____

7. On an order of 60 lamps, 12 were broken when the clerk unpacked them. What percent of the lamps were broken?

Answer _____

8. A leather coat that regularly sells for $400 is on sale for 20% off. What is the sale price of the coat?

Answer _____

PERCENTS

Unit 6 Review

Change each percent to a decimal and then to a fraction. Simplify.

1. 0.4% = _____   11% = _____

2. 35.5% = _____   250% = _____

Change the following to percents.

3. 0.82 = _____   3.45 = _____   0.003 = _____

4. $\frac{7}{8}$ = _____ $\frac{1}{5}$ = _____ $\frac{4}{25}$ = _____

Find each percent or number.

5. 54% of 75   0.9% of 200

6. 6% of 527   7% of 35

7. What percent of 60 is 9?   What percent of 25 is 4?

8. What percent of 55 is 33?   What percent of 18 is 36?

9. 70% of what number is 21?   25% of what number is 17?

Solve.

10. How much will a $160 stereo cost if it is on sale for 30% off?

11. The selling price on toys is a 54% increase over the cost. What is the price of a doll that costs the store $15.25? Round your answer to the nearest cent.

Answer _____   Answer _____

12. What is the interest on a 2-year loan for $1500 at 6.25%?

13. Sales tax is 6.5%. What will the final cost be on a shirt that costs $19.99? Round to the nearest cent.

Answer _____   Answer _____

157

MEASUREMENT
Metric Length

The meter (m) is the basic metric unit of length. A meter can be measured with a meter stick. A baseball bat is about 1 meter long.

A centimeter (cm) is one hundredth of a meter. (Centi means 0.01.) The centimeter is used to measure small lengths. Your small finger is about 1 centimeter across.

A millimeter (mm) is one thousandth of a meter. (Milli means 0.001.) The millimeter is used to measure very small lengths. The head of a pin is about 1 millimeter across.

A kilometer (km) is one thousand meters. (Kilo means 1000.) The kilometer is used to measure long distances. The distance between two cities is measured in kilometers.

> 1 km = 1000 m
> 1 m = 100 cm
> 1 cm = 10 mm

> 1 m = 0.001 km
> 1 cm = 0.01 m
> 1 mm = 0.1 cm

Find: 9.4 m = _____ cm

> To change meters to a smaller unit, multiply.
>
> $$1 \text{ m} = 100 \text{ cm}$$
> $$9.4 \times 100 = 940$$
> $$9.4 \text{ m} = 940 \text{ cm}$$
>
> You will have more centimeters.

Find: 15 m = _____ km

> To change meters to a larger unit, divide.
>
> $$1 \text{ m} = 0.001 \text{ km}$$
> $$15 \div 1000 = 0.015 \text{ km}$$
> $$15 \text{ m} = 0.015 \text{ km}$$
>
> You will have fewer kilometers.

PRACTICE

Circle the best measurement.

	a	*b*
1.	height of a tree	width of a rubber band
	15 m 15 km	4 mm 4m
2.	length of a canoe	height of a step
	5 cm 5 m	30 mm 30cm

Change each measurement to the smaller unit.

	a	*b*	*c*
3.	7 m = _____ cm	15 cm = _____ mm	1200 km = _____ m
4.	225 m = _____ cm = _____ mm	136 km = _____ m	84 m = _____ cm

Change each measurement to the larger unit.

	a	*b*	*c*
5.	120 cm = _____ m	4346 m = _____ km	890 mm = _____ cm
6.	930 mm = _____ cm = _____ m	750 m = _____ km	11 cm = _____ m

PROBLEM SOLVING

Applications

Solve.

1. Burton had a roll of shelf paper 9 meters long. How many centimeters of paper did Burton have?

   Answer _____

2. The Pines have 10 meters between the back of their house and the property line. If they add a room that is 5.75 meters deep, how close will the house come to the property line?

   Answer _____

3. Korinne had 1500 centimeters of fabric. How many meters of fabric does she have?

   Answer _____

4. A tomato plant measured 3 centimeters tall. In 60 days, the plant measured 1.5 meters tall. How much had the plant grown in 60 days?

   Answer _____

5. Mark's best record for the long jump is 3.80 meters. Stan's best jump was 3.45 meters. How much farther was Mark's jump?

   Answer _____

6. Barbara found a snail that was climbing up a tree. When she first saw the snail, it was 3.5 centimeters off the ground. Later that day, the snail was 26 centimeters off the ground. How far had the snail traveled?

   Answer _____

7. Joe bought a piece of material that measured 4.56 meters. He cut the material into 6 equal pieces. How long was each piece?

   Answer _____

8. The Bakers traveled 32,500 meters to see their cousins. How many kilometers did they travel?

   Answer _____

Metric Mass

The word mass is not often used outside the field of science. The common term for mass is weight.

The gram (g) is the basic unit of mass. The gram is used to measure the weight of very light objects. A paper clip weighs about 1 gram.

The kilogram (kg) is one thousand grams. It is used to measure the weight of heavier objects. The weight of a computer would be given in kilograms. Remember, kilo means 1000.

$$1 \text{ kg} = 1000 \text{ g}$$

$$1 \text{ g} = 0.001 \text{ kg}$$

Find: 3.54 kg = _____ g

To change kilograms to a smaller unit, multiply.

$$1 \text{ kg} = 1000 \text{ g}$$
$$3.54 \times 1000 = 3540$$
$$3.54 \text{ kg} = 3540 \text{ g}$$

You will have more grams.

Find: 13.6 g = _____ kg

To change grams to a larger unit, divide.

$$1 \text{ g} = 0.001 \text{ kg}$$
$$13.6 \div 1000 = 0.0136 \text{ kg}$$
$$13.6 \text{ g} = 0.0136 \text{ kg}$$

You will have fewer kilograms.

PRACTICE

Circle the best measurement.

	a	b
1.	weight of a postage stamp	weight of an infant child
	1 g 1 kg	18 g 18 kg
2.	weight of an orange	weight of a bowling ball
	500 g 500 kg	5 g 5 kg

Change each measurement to the smaller unit.

	a	b	c
3.	10 kg = _____ g	4.8 kg = _____ g	0.76 kg = _____ g
4.	0.004 kg = _____ g	1.092 kg = _____ g	305 kg = _____ g

Change each measurement to the larger unit.

	a	b	c
5.	2.8 g = _____ kg	7 g = _____ kg	3094 g = _____ kg
6.	925 g = _____ kg	52.43 g = _____ kg	61 g = _____ kg

PROBLEM SOLVING

Applications

Solve.

1. Art had 2.53 kilograms of nails in a bag. He dropped the bag and spilled some of the nails. He now had 0.176 kilogram of nails. How many kilograms of nails did Art spill?

Answer _____

2. How many grams are in a can of beans that weighs 0.26 kg?

Answer _____

3. A bag of potatoes holds 25 kilograms. How many grams of potatoes are in the bag?

Answer _____

4. A carton of juice weighs 246 grams. How many kilograms does the carton weigh?

Answer _____

5. A paperback book weighs 0.172 kilogram. How many grams does the book weigh?

Answer _____

6. Andy weighed 95 kilograms. After dieting, he weighed 82.8 kilograms. How much weight had Andy lost?

Answer _____

7. A full tin of pepper weighs 43 grams. The empty tin weighs 39.6 grams. How much does the pepper weigh?

Answer _____

8. A package of paper weighs 215 grams. How much will 6 packages of paper weigh?

Answer _____

MEASUREMENT
Metric Capacity

The liter (L) is the basic metric unit of capacity. A liter of liquid will fill a box 10 centimeters on each side. A large bottle of milk holds about 4 liters.

A milliliter (mL) is one thousandth of a liter. It is used to measure very small amounts of liquid. A milliliter of liquid will fill a box 1 centimeter on each side. A container of yogurt holds about 200 milliliters. Remember, milli means 0.001.

$$1 \text{ L} = 1000 \text{ mL}$$

$$1 \text{ mL} = 0.001 \text{ L}$$

Find: 75 L = _____ mL

To change liters to a smaller unit, multiply.

$$1 \text{ L} = 1000 \text{ mL}$$
$$75 \times 1000 = 75,000$$
$$75 \text{ L} = 75,000 \text{ mL}$$

You will have more milliliters.

Find: 1250 mL = _____ L

To change milliliters to a larger unit, divide.

$$1 \text{ mL} = 0.001 \text{ L}$$
$$1250 \div 1000 = 1.250$$
$$1250 \text{ mL} = 1.25 \text{ L}$$

You will have fewer liters.

PRACTICE

Circle the best measurement.

	a	b

1. capacity of a gas tank   capacity of a container of cottage cheese

     60 mL   60 L   750 mL   750 L

2. capacity of a tea cup   capacity of an eye dropper

     300 mL   300 L   50 mL   50 L

Change each measurement to the smaller unit.

     a   b   c

3. 0.7 L = _____ mL   8 L = _____ mL   1.6 L = _____ mL

4. 421 L = _____ mL   3.09 L = _____ mL   0.424 L = _____ mL

Change each measurement to the larger unit.

     a   b   c

5. 8883 mL = _____ L   390.7 mL = _____ L   14 mL = _____ L

6. 12.5 mL = _____ L   208 mL = _____ L   79 mL = _____ L

PROBLEM SOLVING

Applications

Solve.

---

1. Theresa used 375 milliliters of paint to cover 1 cabinet. How much paint does she need to cover 12 cabinets?

Answer _____

2. How many 5-milliliter spoonfuls are in 1000 milliliters of liquid?

Answer _____

---

3. A measuring cup holds 250 milliliters. How many liters does the measuring cup hold?

Answer _____

4. A 1-liter container of medicine costs $234.50. What will be the cost of 3 liters of this medicine?

Answer _____

---

5. Lou's faucet is leaking at the rate of 375 mL an hour. How much water will have leaked out after 4 hours?

Answer _____

6. A can of frozen concentrate orange juice holds 180 milliliters. How much juice is made when three cans of water are added to the concentrate?

Answer _____

---

7. How many liters of medicine can be packaged from a 4500-milliliter container?

Answer _____

8. A container holds 8064 milliliters of liquid. How many liters is this?

Answer _____

PROBLEM-SOLVING STRATEGY
Work Backwards

When you are given an end result, you need to work backwards to solve the problem. First read the problem carefully to find helpful clues. Once you have found the clues, you can work backwards to solve the problem.

EXAMPLE 1
Read the problem.

From 1980–1988 the price of a commuter ticket has increased. From 1980 to 1985 the price of a monthly commuter ticket doubled. In 1986 the price dropped $4 and in 1987 it dropped $6 more. Then in 1988 it doubled again to $32. How much did an original monthly ticket cost in 1980?

List the clues.

Clue 1.   The 1988 price is double the 1987 price.
Clue 2.   The 1987 price is $6 less than the 1986 price.
Clue 3.   The 1986 price is $4 less than the 1985 price.
Clue 4.   The 1985 price is double the 1980 price.

Solve the problem.

1988 price: $32
1987 price: $\frac{1}{2}$ ($32) = $16
1986 price: $16 + $6 = $22
1985 price: $22 + $4 = $26
1980 price: $\frac{1}{2}$ ($26) = $13

The cost of a ticket in 1980 was $13.

EXAMPLE 2
Read the problem.

Mario is driving in the Indianapolis 500-mile automobile race. When he has driven twice as far as he has already driven, he'll be 50 miles from the end. How far has he driven?

List the clues.

Clue.   At 50 miles from the end of the race, he will have driven twice as far as he has already driven.

Solve the problem.

50 miles from the end: $500 - 50 = 450$ miles

$$\frac{1}{2} \times 450 = 225 \text{ miles}$$

Mario has driven 225 miles.

Work backwards to solve the following problems.

1. Dean Morimoto spends 1 hour 20 minutes getting ready for work each day. After leaving his house he walks 10 minutes to a restaurant. There he spends 30 minutes eating breakfast. Then he walks 5 minutes to a bus stop where he catches a bus for a 25-minute ride to work. What time must he get up to reach his office by 9 A.M.?

Answer _____

2. Sunday's high temperature was 3 degrees higher than Saturday's. On Monday, the temperature fell 5 degrees, then rose 7 degrees on Tuesday and 4 more on Wednesday. Then it fell 17 degrees to a record low of 31 on Thursday. What was the temperature on Saturday?

Answer _____

3. Deanne is painting a fence with 94 posts. When she painted 3 times as many as she has already painted, she'll have only 13 more posts to paint. How many has she painted?

Answer _____

4. Lorene took $\frac{1}{4}$ of the pieces of birthday cake. Liz took $\frac{2}{3}$ of what was left. Holly took the remaining 5 pieces. How many pieces of cake were there to start with?

Answer _____

Perimeter of a Triangle

A *triangle* is a figure with three sides. An *equilateral* triangle has three equal sides. An *isosceles* triangle has two equal sides. A *scalene* triangle has no equal sides.

To find the perimeter of a triangle, add the lengths of the sides.

$P = side\ 1 + side\ 2 + side\ 3$

$P = \ \ 58\ \ + \ \ 75\ \ + \ \ 48$

$P = \ \ 181$

The perimeter is 181 inches.

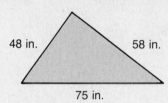

48 in.   58 in.

75 in.

Draw a picture and write in the lengths of each side. Solve.

1. Find the perimeter of an equilateral triangle with one side that measures 9.6 centimeters.

Answer _____

2. Roger needs trim for a triangular-shaped bandana for his costume. The bandana measures 60 centimeters on 2 sides and 90 centimeters. How much trim does he need?

Answer _____

3. How many meters of wire will be required to enclose a triangular-shaped park? The sides measure 50, 45, and 69 meters.

Answer _____

4. Find the perimeter of an equilateral triangle with one side that measures 0.068 millimeters.

Answer _____

5. What is the perimeter of an isosceles triangle that has two sides 8 kilometers long and a third side 6 kilometers long?

Answer _____

6. The sides of a triangular lot measure 130, 120, and 60 meters. What is the perimeter of the lot?

Answer _____

MEASUREMENT

Perimeter of a Rectangle

To find the perimeter of a rectangle, you can use a formula.

Notice that the opposite sides of a rectangle are equal.

The formula, $P = 2l + 2w$, means the perimeter of a rectangle equals 2 times the length plus 2 times the width.

Find the perimeter of this rectangle by using the formula.

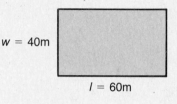

$w = 40m$

$l = 60m$

Write the formula. $P = 2l + 2w$

Substitute the data. $P = (2 \times 60) + (2 \times 40)$

Solve the problem. $P = 120 + 80$

$P = 200$

The perimeter of the rectangle is 200 meters.

Use the formula for perimeter of a rectangle. Solve.

1. Ned wants to fence his garden. It is 15 meters wide and 20 meters long. How much fencing does he need?

Answer _____

2. Harvey Park is 0.25 kilometer wide and 0.175 kilometer long. What is the perimeter of the park?

Answer _____

3. A pillow measures 54 centimeters by 62 centimeters. How much braid is needed to go around the pillow?

Answer _____

4. A desk measures 159 centimeters by 120 centimeters. What is the perimeter of the desk?

Answer _____

5. A farm measures 1.4 kilometers by 1.2 kilometers. What is the perimeter of the farm?

Answer _____

6. A rectangle is twice as long as it is wide. The rectangle is 20.5 millimeters wide. What is the perimeter of the rectangle?

Answer _____

Formula for Area of a Triangle

To find the area (number of square units) of a triangle, you can use a formula.

The formula, $A = \frac{1}{2} bh$, means the area of a triangle equals one half the base times the height.

Find the area of this triangle by using the formula.

Write the formula. $\qquad A = \frac{1}{2} b \times h$

Substitute the data. $\qquad A = \frac{1}{2} \times 12 \times 15$

Solve the problem. $\qquad A = \frac{12 \times 15}{2} = \frac{180}{2} = 90$

The area is 90 *square* centimeters.

Remember, write your answer in *square* units.

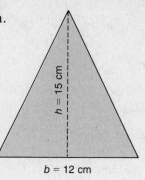

$h = 15$ cm

$b = 12$ cm

Use the formula for area of a triangle. Solve.

1. What is the area of a triangle that has a base of 16 meters and a height of 20 meters?

   Answer _____

2. How much sod will be needed to cover a triangular-shaped park? The base of the park measures 35 meters and the height measures 29 meters.

   Answer _____

3. A triangular-shaped park has one side, the base, measuring 120 meters and one side, the height, measuring 80 meters. What is the area of the park?

   Answer _____

4. A new highway cut off part of the town and formed a triangle. The base of the property measures 30 kilometers and one side, the height, measures 25 kilometers. What is the area of this part of the town?

   Answer _____

5. The top of the front wall of a barn forms a triangle. The triangle is 3.5 meters tall and 10 meters wide. What is the area of this part of the wall?

   Answer _____

6. School banners are shaped like triangles. The base of the triangle is 11 inches. The height of the triangle is 25 inches. How much material is needed to make each banner?

   Answer _____

MEASUREMENT

Formula for Area of a Rectangle

To find the area (number of square units) of a rectangle, you can use a formula.

The formula, $A = lw$, means the area of a rectangle equals the length times the width.

Find the area of this rectangle by using the formula.

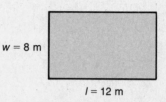

Write the formula. $A = l \times w$

Substitute the data. $A = 12 \times 8$

Solve the problem. $A = 96$

The area of the rectangle is 96 square meters.

Remember, write your answer in *square* units.

Use the formula for area of a rectangle. Solve.

1. How many square meters are in the floor area of a room that measures 8 meters by 6.9 meters?

    Answer _____

2. What is the area of a window that measures 16 inches by 24 inches?

    Answer _____

3. A pillow measures 62 centimeters by 40 centimeters. How much fabric will be needed to cover one side of the pillow?

    Answer _____

4. How much sod is needed to cover a yard that measures 20 meters by 15 meters?

    Answer _____

5. What is the area of a concrete walk that measures 100 meters by 1.5 meters?

    Answer _____

6. How much linoleum is needed to cover a floor that measures 54 feet by 42 feet?

    Answer _____

7. How many square inches of glass are needed to cover a picture that measures 8 inches by 10 inches?

    Answer _____

8. How much carpeting is needed to cover a floor that measures 3 meters by 4.4 meters?

    Answer _____

MEASUREMENT
Formula for Volume of a Rectangular Solid

To find the volume (number of cubic units) of a rectangular solid, you can use a formula.

The formula, $V = lwh$, means the volume of a rectangular solid equals the length times the width times the height.

Find the volume of this rectangular solid.

Write the formula. $V = l \times w \times h$
Substitute the data. $V = 10 \times 15 \times 4$
Solve the problem. $V = 600$

h = 4 m
l = 10 m
w = 15 m

The volume is 600 *cubic* meters.

Remember, write your answer in *cubic* units.

Use the formula for volume of a rectangular solid. Solve.

1. How many cubic meters of concrete are needed for a driveway that measures 2 meters wide, 27 meters long, and 0.1 meters thick?

Answer _____

2. A classroom measures 14 meters by 7 meters by 3.5 meters. How many cubic meters of air are in the room?

Answer _____

3. How much dirt was removed from a hole that measures 80 meters long, 54 meters wide, and 4 meters deep?

Answer _____

4. What is the volume of a basement that measures 18 meters by 13.4 meters by 4.2 meters?

Answer _____

5. How many liters of gas are in a tank that measures 55 centimeters by 35 centimeters by 31 centimeters?

Answer _____

6. What is the volume of water in a tank that measures 10 meters by 12 meters by 8 meters?

Answer _____

7. How much soil is needed to fill a box that is 12 inches by 19 inches by 18 inches?

Answer _____

8. What is the volume of a bin that measures 5 meters by 2 meters by 3.5 meters?

Answer _____

MEASUREMENT

Formula for Circumference of a Circle

Circumference is the distance around a circle. To find the circumference, you can use a formula.

The formula, $C = \pi d$, means the circumference equals pi times the diameter. This formula uses the symbol π, or pi. We use 3.14 for the value of π. Notice in the drawing that the diameter is a straight line across the circle and through the center.

You can also use the formula, $C = 2\pi r$, which means the circumference equals two times pi times the radius. Notice in the drawing that the *radius* is a straight line from the center of the circle to the circumference. It is one half of the diameter, so $2r = d$.

Find the circumference of this circle.

Write the formula. $C = \pi d$

Substitute the data. $C = 3.14 \times 12$

Solve the problem. $C = 37.68$

diameter = 12 mm

radius = 6 mm

The circumference of the circle is 37.68 millimeters.

Use the formula for the circumference of a circle. Solve.

1. What is the circumference of a circle that has a diameter of 7 meters?

Answer _____

2. The wheel of a bicycle has a radius of 35 centimeters. What is the circumference of the wheel?

Answer _____

3. What is the circumference of a circle with a radius of 175 millimeters?

Answer _____

4. What is the circumference of a circle with a diameter of 3.5 centimeters?

Answer _____

5. At the Alamo in San Antonio a circular flower garden encloses the star of Texas. The garden has a radius of 3.15 meters. How much fence is needed to enclose the garden? Round your answer to the nearest hundredth.

Answer _____

6. The diameter of the inside of the ring of a basketball goal is 45.71 centimeters. What is the circumference of the ring? Round your answer to the nearest hundredth.

Answer _____

Formula for Area of a Circle

To find the area of a circle, you can use a formula.

The formula, $A = \pi r^2$, means the area of a circle is equal to pi times the radius squared, or the radius times the radius.

Find the area of this circle.

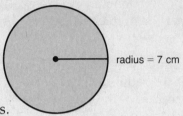

radius = 7 cm

Write the formula. $A = \pi \ (r \times r)$

Substitute the data. $A = 3.14 \ (7 \times 7)$

Solve the problem. $A = 3.14 \times 49 = 153.86$

The area of the circle is 153.86 square centimeters.

Remember, write your answer in *square* units.

Use the formula for area of a circle. Solve.

---

1. Find the area of a circle that has a radius of 3.5 meters.

Answer _____

2. What is the area of a circle that has a diameter of 28 meters? (Hint: the radius is one half the diameter.)

Answer _____

---

3. A bandstand in the shape of a circle is to be 7 meters across. How many square meters of flooring will be required? (Hint: the radius is one half the diameter.) Round your answer to the nearest hundredth.

Answer _____

4. The circular canvas net used by fire fighters has a radius of 2.1 meters. What is the area of the net?

Answer _____

---

5. A gallon of paint covers 56 square meters. How many gallons of paint will be needed to cover a circular floor that is 22 meters in diameter? (Hint: the radius is one half the diameter.)

Answer _____

6. The top of a piston is a circle. The top of an auto piston has a radius of 4.2 centimeters. What is the area of the top of the piston? Round your answer to the nearest hundredth.

Answer _____

MEASUREMENT
Formula for Volume of a Cylinder

To find the volume (cubic units) of a cylinder, you can use a formula.

The formula, $V = \pi r^2 h$, means the volume of a cylinder equals pi times the radius squared times the height.

Remember, diameter = $2r$.

Find the volume of this cylinder.

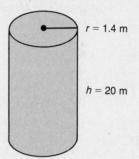

Write the formula. $V = \pi\ (r \times r) \times h$

Substitute the data. $V = 3.14 \times (1.4 \times 1.4) \times 20$

Solve the problem. $V = 123.088$

The volume of the cylinder is 123.088 cubic meters.

Remember, write your answer in *cubic* units.

Use the formula for volume of a cylinder. Solve.

---

1. A cylindrical water tank has a diameter of 2.8 meters and is 6.5 meters high. What is the volume of the tank? Round your answer to the nearest hundredth.

Answer _____

2. The inside radius of a pipe is 35 centimeters. One section of pipe is 6 meters long. How much water will this piece of pipe hold?

Answer _____

---

3. A test tube has a radius of 1 centimeter and is 3.5 centimeters long. How much liquid can the test tube hold?

Answer _____

4. A storage tank has a radius of 5.25 meters and a height of 12 meters. How much liquid can the storage tank hold? Round your answer to the nearest hundredth.

Answer _____

---

5. The tank of a gasoline truck has a radius of 1.75 meters and is 7 meters long. What is the volume of the tank? Round your answer to the nearest hundredth.

Answer _____

6. A cylinder-shaped container is 9 centimeters in diameter and 15 centimeters in height. How much liquid will this container hold? Round your answer to the nearest hundredth.

Answer _____

PROBLEM-SOLVING STRATEGY

Use a Line Graph

A line graph may show how something can change over a period of time. The dots represent data (factual information). Lines connect the dots and show increase or decrease over time. Use a line graph to answer questions about the data.

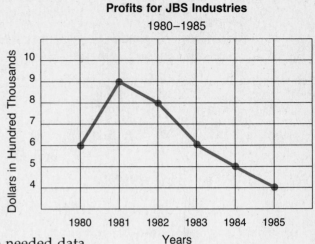

Profits for JBS Industries
1980–1985

This graph shows the profits of JBS Industries over a six-year period from 1980 to 1985.

Read the problem.
What was the increase in profits from 1980–1981?

Read the graph for facts.
Use these steps when finding the needed data.
1. Locate the year on the horizontal scale along the bottom of the graph.
2. For each year, move up the line to the dot and back to the vertical scale along the side of the graph.
3. Write down the data.
   Data: In 1980 JBS Industries earned $600,000. (6 × 100,000)
   In 1981 JBS Industries earned $900,000. (9 × 100,000)

Solve the problem.
$900,000 − $600,000 = $300,000

The profits for JBS Industries increased by $300,000.

Solve.

1. The line graph shows the profits for JBS Industries for how many years? Answer _____	**2.** What is the dollar value of each number at the left of the graph? Answer _____
3. What was the decrease in profits between 1981 and 1982? Answer _____	**4.** What was the decrease in profits between 1983 and 1984? Answer _____
5. What was the decrease in profits between 1981 and 1985? Answer _____	**6.** Between which two successive years did the profits decrease the most? How much was the decrease? Answer _____

Solve. Use the line graph below.

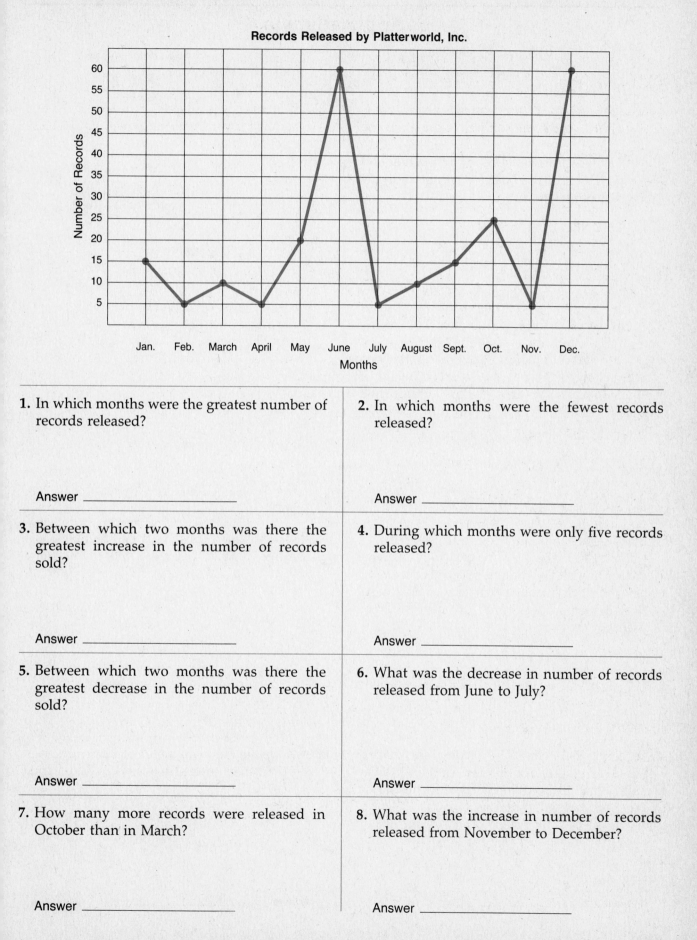

Records Released by Platterworld, Inc.

1. In which months were the greatest number of records released?

   Answer _____

2. In which months were the fewest records released?

   Answer _____

3. Between which two months was there the greatest increase in the number of records sold?

   Answer _____

4. During which months were only five records released?

   Answer _____

5. Between which two months was there the greatest decrease in the number of records sold?

   Answer _____

6. What was the decrease in number of records released from June to July?

   Answer _____

7. How many more records were released in October than in March?

   Answer _____

8. What was the increase in number of records released from November to December?

   Answer _____

PROBLEM SOLVING

Applications

Solve.

1. A dollhouse sits on a rectangular board that is 36 inches by 24 inches. Fencing for the dollhouse costs $0.15 per inch. How much will it cost to build a fence around the board? (Hint: first find the perimeter of the board. Then multiply by the cost of the fencing.)

Answer _____

2. The diameter of a bicycle tire is 66 centimeters. How far forward will the wheel have moved after 6 complete turns?

Answer _____

3. A kite is made up of 4 triangles. Two of the triangles have a base of 30 centimeters and a height of 20 centimeters. The other 2 triangles have a base of 30 centimeters and a height of 60 centimeters. What is the total area of the kite?

Answer _____

4. A rectangle measures 12 inches by 9 inches. A triangle has a base of 12 inches and a height of 17 inches. Compute the area of each. Which figure has the greater area?

Answer _____

5. Blake needs to build a frame for his piece of stained glass. The piece is a circle with a diameter of 18 inches. Framing material costs $0.10 per inch. How much will the frame cost? Round your answer to the nearest hundredth.

Answer _____

6. A sandbox measures 5 feet by 4 feet by 1 foot. Sand costs $0.75 per cubic foot. How much will it cost to fill the sandbox?

Answer _____

7. Frozen yogurt comes in a container shaped like a rectangular solid. The container measures 6 inches by 10 inches by 4 inches. How much yogurt will it hold?

Answer _____

8. A pie plate has a diameter of 22 centimeters. How much counter space is needed to display 8 pies?

Answer _____

MEASUREMENT

Unit 7 Review

Change each measurement to the smaller unit.

	a	b	c

1. 346 m = _____ cm   9.5 kg = _____ g   0.4 L = _____ mL

Change each measurement to the larger unit.

	a	b	c

2. 33.8 mL = _____ L   2 g = _____ kg   0.71 mm = _____ cm = _____ m

Solve.

3. The path Jorge jogs forms a rectangle. The sides of the rectangle are 6 blocks and 9 blocks. How many blocks does Jorge jog?

Answer _____

4. The sail on a boat is shaped like a triangle. It is 7 yards tall and is 4 yards along the base. How many yards of material are needed to make the sail?

Answer _____

5. Karla made a puzzle that formed a triangle. The sides of the puzzle were 5 inches, 13 inches, and 12 inches. How many inches of wood does Karla need to make a frame for the puzzle?

Answer _____

6. A rectangular-shaped toolbox is 24 inches long, 12 inches wide, and 8 inches tall. What is the volume of the toolbox?

Answer _____

7. A circular-shaped garden has a radius of 12 feet. How much plastic edging will be needed to go around the garden?

Answer _____

8. A window measures 1.5 meters by 2.3 meters. How much glass is needed for the window?

Answer _____

SOLVING AND USING EQUATIONS

Finding Missing Addends and Missing Factors

When you add two numbers, the numbers that you add are called **addends.** When you multiply two numbers, the numbers you multiply are called **factors.** An **equation** is a statement that two quantities are equal. Letters are often used to stand for unknown or **missing** addends or factors. For example, in the equation $x + 3 = 11$, the letter x stands for the unknown or missing addend that you would add to 3 to get 11. In the equation $3a = 27$, the letter a stands for the unknown or missing factor you would multiply by to get 27.

Remember, $3a$, $3 \cdot a$, and $3(a)$ mean "3 times a".

To find a missing addend, subtract the known addend from both sides of the equation. To find a missing factor, divide both sides of the equation by the known factor. Check by substituting the answer into the original equation.

Solve for a missing addend: $x + 3 = 11$

$$x + 3 = 11$$
$$x + 3 - 3 = 11 - 3 \quad \text{Subtract 3.}$$
$$x = 8$$

Check: $x + 3 = 11$
$$8 + 3 = 11$$

Solve for a missing factor: $3a = 27$

$$3a = 27$$
$$\frac{3a}{3} = \frac{27}{3} \quad \text{Divide by 3.}$$
$$a = 9$$

Check: $3a = 27$
$$3(9) = 27$$

GUIDED PRACTICE

Solve. Check.

	a	b	c	d
1.	$x + 7 = 10$ $x + 7 - 7 = 10 - 7$ $x = 3$ Check: $x + 7 = 10$ $3 + 7 = 10$	$5x = 20$ $\frac{5x}{5} = \frac{20}{5}$ $x =$	$5k = 36$ $\frac{5k}{5} = \frac{36}{5}$ $k =$	$y + 5 = 15$ $y + 5 - 5 = 15 - 5$ $y =$
2.	$10m = 80$ $\frac{10m}{10} = \frac{80}{10}$ $m =$	$n + 12 = 29$ $n + 12 - 12 = 29 - 12$ $n =$	$8x = 96$ $\frac{8x}{8} = \frac{96}{8}$ $x =$	$p + 7 = 35\frac{1}{2}$ $p + 7 - 7 = 35\frac{1}{2} - 7$ $p =$
3.	$9\frac{1}{3} + x = 30\frac{2}{3}$ $9\frac{1}{3} + x - 9\frac{1}{3} = 30\frac{2}{3} - 9\frac{1}{3}$ $x =$	$25x = 200$ $\frac{25x}{25} = \frac{200}{25}$ $x =$	$7k = 91$ $\frac{7k}{7} = \frac{91}{7}$ $k =$	$40 + x = 93$ $40 + x - 40 = 93 - 40$ $x =$

Solve. Simplify.

a	b	c	d
1. $x + \frac{1}{4} = \frac{3}{4}$	$9n = 315$	$3x = 7$	$1\frac{2}{3} + x = 5$
2. $k + 15 = 40$	$16x = 90$	$m + 6 = 7\frac{3}{8}$	$12k = 156$
3. $5x = 37$	$81 + k = 92$	$x + 1\frac{1}{2} = 2\frac{1}{4}$	$11x = 572$
4. $6k = 70$	$x + 46 = 75$	$3x = 261$	$1\frac{1}{8} + x = 6\frac{1}{4}$
5. $21n = 105$	$21n = 45$	$63 + x = 79$	$n + 7\frac{1}{2} = 15$
6. $13 + x = 81$	$5x = 43$	$12x = 144$	$x + 102 = 200$

→ **MIXED PRACTICE**

Find each answer.

a	b	c	d
1. $\begin{array}{r} 437 \\ \times\ 27 \\ \hline \end{array}$	$\begin{array}{r} 1296 \\ \times\ \ \ 54 \\ \hline \end{array}$	$\begin{array}{r} 3247 \\ \times\ 129 \\ \hline \end{array}$	$\begin{array}{r} 4069 \\ \times\ 205 \\ \hline \end{array}$
2. $16\overline{)992}$	$26\overline{)2075}$	$36\overline{)1476}$	$25\overline{)9250}$

SOLVING AND USING EQUATIONS

Solving Equations

Some equations require two or more operations to solve for the unknown.

Remember,
- add, subtract, multiply, or divide the **same** number on **both** sides of the equal sign.
- you cannot divide by zero.

Solve: $5x - 2 = 38$

$$5x - 2 = 38$$
$$5x - 2 + 2 = 38 + 2 \quad \text{Add 2.}$$
$$5x = 40$$
$$\frac{5x}{5} = \frac{40}{5} \quad \text{Divide by 5.}$$
$$x = 8$$

Check: $5x - 2 = 38$
$5(8) - 2 = 38$
$40 - 2 = 38$
$38 = 38$

GUIDED PRACTICE

Solve. Check.

a	b	c

1.

a
$$7x + 9 = 51$$
$$7x + 9 - 9 = 51 - 9$$
$$7x = 42$$
$$\frac{7x}{7} = \frac{42}{7}$$
$$x = 6$$

Check: $7x + 9 = 51$
$7(6) + 9 = 51$
$42 + 9 = 51$
$51 = 51$

b
$$3x - 6 = 30$$
$$3x - 6 + 6 = 30 + 6$$

c
$$6x + \frac{1}{5} = 65$$
$$6x + \frac{1}{5} - \frac{1}{5} = 65 - \frac{1}{5}$$

2.

$$10x + 9 = 59$$
$$10x + 9 - 9 = 59 - 9$$

$$4x - \frac{1}{2} = 270$$
$$4x - \frac{1}{2} + \frac{1}{2} = 270 + \frac{1}{2}$$

$$5x - 10 = 60$$
$$5x - 10 + 10 = 60 + 10$$

3.

$$8x + 32 = 200$$
$$8x + 32 - 32 = 200 - 32$$

$$9x + 11 = 74$$
$$9x + 11 - 11 = 74 - 11$$

$$13x + 13 = 143$$
$$13x + 13 - 13 = 143 - 13$$

Solve. Check.

a	*b*	*c*
1. $12x + 7 = 67$	$6x + 1 = 4$	$2x + 16 = 32$

2. $7x - 5\frac{1}{2} = 12$ $2x + 19 = 29$ $4x + 7 = 10$

3. $9x + 92 = 128$ $8x - 35 = 37$ $5x - \frac{1}{2} = \frac{1}{2}$

4. $8x + \frac{1}{3} = \frac{2}{3}$ $2x - \frac{1}{5} = \frac{2}{5}$ $10x - 8 = 32$

➡ **MIXED PRACTICE**

Find each answer.

a	*b*	*c*	*d*
1. 1735 967 +8446	8102 − 906	231 × 18	62)756
2. 1.03 14.708 + 9.317	4.002 −1.9364	2.16 × 0.8	0.7)1.54

SOLVING AND USING EQUATIONS

Collecting Terms in Equations

In the equation $2x + 3x = 15$, the two addends, called terms, both contain the unknown (x) as a factor. They are **like** terms. These two terms, $2x$ and $3x$, can be added to get the single term $5x$ because they are like terms. This process is called collecting like terms. Check by substitution.

Collect like terms and solve: $2x + 3x = 15$

$$2x + 3x = 15$$
$$5x = 15 \quad \text{Collect like terms.}$$
$$\frac{5x}{5} = 15 \quad \text{Divide by 5.}$$
$$x = 3$$

Check: $2(3) + 3(3) = 15$
$$6 + 9 = 15$$
$$15 = 15$$

Collect like terms and solve: $9x - 3x = 24$

$$9x - 3x = 24$$
$$6x = 24 \quad \text{Collect like terms.}$$
$$\frac{6x}{6} = 24 \quad \text{Divide by 6.}$$
$$x = 4$$

Check: $9(4) - 3(4) = 24$
$$36 - 12 = 24$$
$$24 = 24$$

GUIDED PRACTICE

Solve. Check.

a $\qquad\qquad\qquad\qquad$ b $\qquad\qquad\qquad\qquad$ c

1. $\frac{1}{2}x + 4x = 45$ $\qquad\qquad$ $3x + 7x = 55$ $\qquad\qquad$ $23x + 27x = 100$

$4\frac{1}{2}x = 45$ $\qquad\qquad\qquad\quad$ $10x = 55$ $\qquad\qquad\qquad\quad$ $50x = 100$

$\frac{9}{2}x = 45$

$x = \overset{5}{\cancel{45}}\left(\frac{2}{\cancel{9}}\right)$

$x = 10$

Check: $\frac{1}{2}(10) + 4(10) = 45$
$$5 + 40 = 45$$

2. $19x - 7x = 84$ $\qquad\qquad$ $16x + 12x = 84$ $\qquad\qquad$ $35x - 2\frac{1}{2}x = 65$

$12x = 84$ $\qquad\qquad\qquad\quad$ $28x = 84$ $\qquad\qquad\qquad\quad$ $32\frac{1}{2}x = 65$

3. $10x + \frac{2}{5}x = 156$ $\qquad\qquad$ $17x + 12x = 87$ $\qquad\qquad$ $12x + 5x = 51$

$10\frac{2}{5}x = 156$ $\qquad\qquad\qquad\quad$ $29x = 87$ $\qquad\qquad\qquad\quad$ $17x = 51$

Collect like terms. Then solve. Check.

a	b	c
1. $6x - 2x = 4$	$7x - 3x = 20$	$9x + 11x = 80$
2. $15x - 3x = 144$	$20x - 9x = 77$	$5x + 3x = 6$
3. $2x + 14x = 8$	$5x - x = 24$	$7x + 7x = 70$
4. $2x + 5x = 1\frac{1}{2}$	$9x - 4x = 5$	$30x - 10x = 1\frac{1}{3}$
5. $18x - 15x = 39$	$3\frac{1}{2}x - 1\frac{1}{2}x = 18$	$\frac{1}{3}x + \frac{2}{3}x = 10$
6. $7x + x = 72$	$15x - 5x = 20$	$13x - x = 28$

MIXED PRACTICE

Find each answer.

a	b	c
1. 17% of 50	$33\frac{1}{3}$% of 99	1.5% of 44

Solving Equations with an Unknown on Both Sides

To solve an equation in which an unknown is on both sides of the equation, use the methods you have learned to move all the unknowns to the same side of the equation. Combine like terms and solve. Check by substitution.

Solve: $7x + 3 = 15 + 5x$

$$7x + 3 = 15 + 5x$$
$$7x + 3 - 3 = 15 + 5x - 3 \quad \text{Subtract 3 from both sides.}$$
$$7x = 12 + 5x$$
$$7x - 5x = 12 + 5x - 5x \quad \text{Subtract 5x from both sides.}$$
$$2x = 12$$
$$\frac{2x}{2} = \frac{12}{2} \quad \text{Divide both sides by 2.}$$
$$x = 6$$

Check: $7(6) + 3 = 15 + 5(6)$
$42 + 3 = 15 + 30$
$45 = 45$

GUIDED PRACTICE

Solve.

	a	*b*	*c*

1.

a:
$$9x + 6 = 4x + 36$$
$$9x + 6 - 6 = 4x + 36 - 6$$
$$9x - 4x = 4x + 30 - 4x$$
$$5x = 30$$
$$\frac{5x}{5} = \frac{30}{5}$$
$$x = 6$$

Check: $9(6) + 6 = 4(6) + 36$
$54 + 6 = 24 + 36$
$60 = 60$

b:
$$5x - 9 = 1 + 3x$$
$$5x - 9 + 9 = 1 + 9 + 3x$$
$$5x - 3x = 10 + 3x - 3x$$

c:
$$10x = 8 + 2x$$
$$10x - 2x = 8 + 2x - 2x$$

2.

a:
$$5x - 3 = 17 - 5x$$
$$5x - 3 + 3 = 17 - 5x + 3$$
$$5x + 5x = 20 - 5x + 5x$$

b:
$$3x + 1 = 41 - 2x$$
$$3x + 1 - 1 = 41 - 2x - 1$$
$$3x + 2x = 40 - 2x + 2x$$

c:
$$11x - 5 = 45 + 6x$$
$$11x - 5 + 5 = 45 + 6x + 5$$
$$11x - 6x = 50 + 6x - 6x$$

3.

a:
$$7x + 16 = 32 - x$$
$$7x + 16 - 16 = 32 - x - 16$$
$$7x + x = 16 - x + x$$

b:
$$2x = 7 + x$$
$$2x - x = 7 + x - x$$

c:
$$20x - 13 = 15x + 62$$
$$20x - 13 + 13 = 15x + 62 + 13$$
$$20x - 15x = 15x + 75 - 15x$$

Solve.

	a	*b*	*c*
1.	$25x + 9 = 209 + 5x$	$7x - 10 = 14 - x$	$16x + 5 = 59 + 7x$
2.	$3x - 12 = 2x + 5$	$10x + 6 = 7 + 9x$	$23x + 100 = 725 - 2x$
3.	$15x + 3 = 13x + 13$	$6x + 8 = 98 + 3x$	$3x + 2 = 50 - 3x$
4.	$15x + 7 = 107 + 5x$	$8x - 6 = 42 + 6x$	$6x = 27 + 3x$

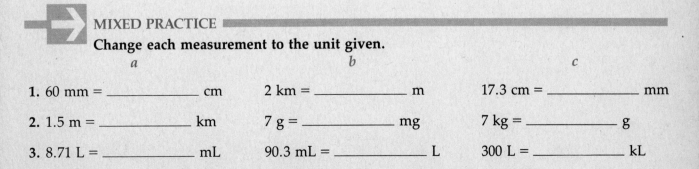

MIXED PRACTICE

Change each measurement to the unit given.

	a	*b*	*c*
1.	60 mm = _____ cm	2 km = _____ m	17.3 cm = _____ mm
2.	1.5 m = _____ km	7 g = _____ mg	7 kg = _____ g
3.	8.71 L = _____ mL	90.3 mL = _____ L	300 L = _____ kL

SOLVING AND USING EQUATIONS

Using Equations to Solve Problems

Many kinds of problems can be solved by deciding what is unknown, writing an equation, and solving. Check your answer in the original problem.

Jack and Lisa together have $250. Jack has $10 more than 3 times as much as Lisa. How much does each have?

Decide what is unknown.
$\quad$ Let x = Lisa's money
$\quad 3x + 10$ = Jack's money

Write an equation.
$\quad$ Lisa's money + Jack's money = $250
$\quad\quad x + 3x + 10 = \$250$

Solve.
$$x + 3x + 10 = 250$$
$$4x + 10 = 250$$
$$4x + 10 - 10 = 250 - 10$$
$$4x = 240$$
$$\frac{4x}{4} = \frac{240}{4}$$
$$x = 60$$

Answer:
$\quad x = \$60 \quad$ Lisa's money
$\quad 3x + 10 = \$190 \quad$ Jack's money

GUIDED PRACTICE

Write an equation and solve.

1. If 24 is added to a certain number, the result is 52. What is the number?

Let n = the number.

$$n + 24 = 52$$
$$n + 24 - 24 = 52 - 24$$
$$n = 28$$

Answer _____ 28 _____

2. Terry has $25 more than twice as much money as Tom has. Together they have $145. How much money does each have?

$\quad$ Let x = Tom's money.
Let $2x + 25$ = Terry's money.

$$x + 2x + 25 = 145$$

Tom _____

Terry _____

3. Pat has three times as much money as does Jerry. Together they have $12.80. How much does each have?

$\quad$ Let x = Jerry's money.
Let $3x$ = Pat's money.

Jerry _____

Pat _____

4. Louisa is thinking of a number. Five times that number equals 240. What is the number?

Let n = the number.

Answer _____

Write an equation and solve.

1. Jane has five times as much money as does Helen. Together they have $42. How much does each have?

   Jane _____

   Helen _____

2. Ruth has $15 more than Julia. Together they have $55. How much does each have?

   Ruth _____

   Julia _____

3. One package contains 20 envelopes more than a second package. Together they contain 80 envelopes. How many envelopes does each package contain?

   1st package _____

   2nd package _____

4. One lot contains 50 square feet less than a second lot. Together the lots contain 200 square feet. How many square feet are there in each lot?

   1st lot _____

   2nd lot _____

5. Joe's mother is 3 times as old as Joe. The sum of their ages is 72 years. How old is each?

   Mother _____

   Joe _____

6. Frank's father is 2 years less than three times as old as Frank. The sum of their ages is 78 years. How old is each?

   Father _____

   Frank _____

7. Three numbers add up to 180. The second number is twice the first, and the third is three times the first. What is each number?
   Let n = the first number.

   1st _____

   2nd _____

   3rd _____

8. Three numbers add up to 140. The second number is twice the first, and the third number is twice the second. What are the three numbers?

   1st _____

   2nd _____

   3rd _____

SOLVING AND USING EQUATIONS

Ratios

A ratio is a fraction used to compare two quantities. For example, if a baseball player gets 3 hits for every 6 times at bat, the ratio of hits to times at bat is $\frac{3}{6}$ or $\frac{1}{2}$. The ratio of times at bat to hits is $\frac{6}{3}$ or $\frac{2}{1}$.

PRACTICE

Write a fraction for each ratio.

a   *b*

1. The ratio of inches in a foot to inches in a yard

   Ratio: _____$\frac{12}{36}$ or $\frac{1}{3}$_____

   The ratio of hours in a day to hours in a week

   Ratio: _____

2. The ratio of cups in a pint to cups in a quart

   Ratio: _____

   The ratio of 3 apples on a table to 6 apples in a bowl

   Ratio: _____

3. The ratio of cents in a quarter to cents in a dollar

   Ratio: _____

   The ratio of 8 men to 10 women

   Ratio: _____

4. The ratio of cents in a half dollar to cents in a dime

   Ratio: _____

   The ratio of 10 women to 8 men

   Ratio: _____

5. The ratio of 17 "yes" votes to 20 "no" votes

   Ratio: _____

   The ratio of 5 wall outlets to 3 wall switches

   Ratio: _____

6. The ratio of the number of days in December to the number of days in January

   Ratio: _____

   The ratio of 8 hours asleep to 16 hours awake

   Ratio: _____

7. The ratio of minutes in an hour to minutes in a half hour

   Ratio: _____

   The ratio of 8 chicken legs to 8 chicken wings

   Ratio: _____

SOLVING AND USING EQUATIONS

Ratios in Measurement

Here you will need to use what you know about finding areas of squares and rectangles, and volumes of rectangular solids to form ratios.

$$A = lw$$
$$V = lwh$$

The ratio of the sides of two squares is $\frac{1}{2}$. What is the ratio of the areas?

$$\frac{\text{Area A}}{\text{Area B}} = \frac{lw}{lw} = \frac{1 \times 1}{2 \times 2} = \frac{1}{4}$$

square A ▢ 1   square B ▦ 2

The ratio of the two areas is $\frac{1}{4}$.

The area of square B is 4 times greater than the area of square A.

Or, the area of square A is $\frac{1}{4}$ the area of square B.

PRACTICE

Solve.

1. The ratio of the sides of two squares is $\frac{1}{3}$. What is the ratio of the areas?

   Answer _____

2. The ratio of the sides of two squares is $\frac{1}{2.5}$. What is the ratio of the areas?

   Answer _____

3. The ratio of the sides of two cubes is $\frac{1}{5}$. What is the ratio of the volumes?

   Answer _____

4. The ratio of the sides of two cubes is $\frac{3}{4}$. What is the ratio of the volumes?

   Answer _____

5. The ratio of the sides of two cubes is $\frac{2}{3}$. What is the ratio of the volumes?

   Answer _____

6. The ratio of the sides of two squares is $\frac{10}{1}$. What is the ratio of the areas?

   Answer _____

Proportions

A proportion is an equation stating that two ratios are equal. For example, if you mix 2 gallons of red paint with 3 gallons of white, the ratio of red to white is $\frac{2}{3}$. If you mix 4 gallons of red with 6 gallons of white, the ratio of red to white is $\frac{4}{6}$. The shades of pink for the two mixtures are the same, because $\frac{2}{3} = \frac{4}{6}$ is a true proportion.

A quick way to check that two ratios are equal is to cross-multiply. Write the ratios side by side and draw double-pointed arrows that cross. Multiply the pairs of numbers and see if you get the same result both times. It does not matter which pair of numbers is multiplied first. Look at this example.

$$\frac{2}{3} \diagtimes \frac{4}{6}$$

$$(2)(6) = 12 \text{ and } (3)(4) = 12$$

Since you get the same result (12), then $\frac{2}{3} = \frac{4}{6}$. If you get different results when you cross-multiply, the ratios are not equal.

PRACTICE ———————————————————————————————

Use cross-multiplying to tell whether the proportion is true or false. Write "true" or "false".

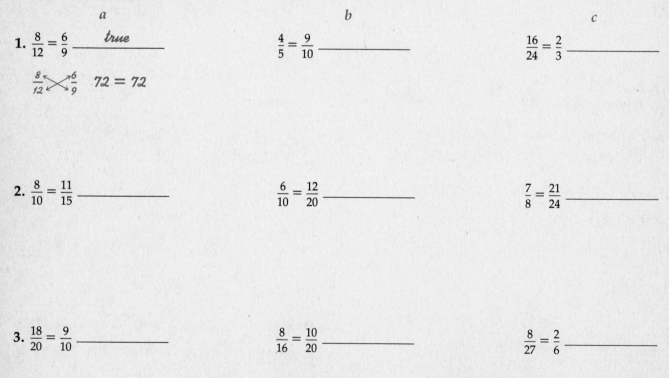

	a	*b*	*c*
1.	$\frac{8}{12} = \frac{6}{9}$ _true_	$\frac{4}{5} = \frac{9}{10}$	$\frac{16}{24} = \frac{2}{3}$
	$\frac{8}{12} \diagtimes \frac{6}{9}$ $72 = 72$		
2.	$\frac{8}{10} = \frac{11}{15}$	$\frac{6}{10} = \frac{12}{20}$	$\frac{7}{8} = \frac{21}{24}$
3.	$\frac{18}{20} = \frac{9}{10}$	$\frac{8}{16} = \frac{10}{20}$	$\frac{8}{27} = \frac{2}{6}$

SOLVING AND USING EQUATIONS

Solving Proportions with an Unknown

To solve a proportion that contains an unknown, cross-multiply and solve the resulting equation. Check your answer in the original problem.

Solve: $\frac{4}{6} = \frac{x}{18}$

$$\frac{4}{6} = \frac{x}{18}$$

$(4)(18) = 6x$ Cross-multiply.
$72 = 6x$
$\frac{72}{6} = \frac{6x}{6}$ Divide.
$12 = x$

Check: $\frac{4}{6} = \frac{x}{18}$
$\frac{4}{6} = \frac{12}{18}$
$4(18) = 6(12)$
$72 = 72$

PRACTICE

Solve.

a	b	c

1. $\frac{5}{10} = \frac{10}{x}$ $\qquad$ $\frac{5}{12} = \frac{10}{x}$ $\qquad$ $\frac{25}{3} = \frac{100}{x}$

$5x = (10)(10)$
$5x = 100$
$\frac{5x}{5} = \frac{100}{5}$
$x = 20$

2. $\frac{7}{9} = \frac{21}{x}$ $\qquad$ $\frac{4}{15} = \frac{8}{x}$ $\qquad$ $\frac{5}{8} = \frac{x}{16}$

3. $\frac{5}{x} = \frac{15}{9}$ $\qquad$ $\frac{20}{x} = \frac{5}{4}$ $\qquad$ $\frac{x}{12} = \frac{25}{3}$

4. $\frac{12}{4} = \frac{x}{7}$ $\qquad$ $\frac{x}{4} = \frac{5}{10}$ $\qquad$ $\frac{x}{20} = \frac{4}{5}$

SOLVING AND USING EQUATIONS

Using Proportions to Solve Problems

There are many problems that you can solve by setting up and solving a proportion.

If Paul can walk 15 miles in 6 hours, how far can he walk, at the same rate, in 8 hours?

Let x = number of miles he can walk in 8 hours.

$$\frac{15}{6} = \frac{x}{8}$$

$$6x = 120$$

$$x = 20 \qquad \text{He can walk 20 miles in 8 hours.}$$

Check: $\frac{15}{6} = \frac{x}{8}$

$$\frac{15}{6} = \frac{20}{8}$$

$$8(15) = 6(20)$$

$$120 = 120$$

PRACTICE

Solve.

1. Lauren bought 6 feet of wire for $21.66. How much would 10 feet of the same wire have cost?

Let x = the cost of 10 feet of wire.

$$\frac{21.66}{6} = \frac{x}{10}$$

$$6x = 216.60$$

$$x = \$36.10$$

Answer _____$36.10_____

2. Kim bought a 14-ounce bottle of catsup for 77¢. At the same rate, how much would a 20-ounce bottle cost?

Answer _____

3. Joe used $3.20 worth of gasoline to drive 68 miles. How many dollar and cents worth of gas will he use to drive 85 miles?

Answer _____

4. The tax on a piece of property valued at $20,000 was $264. At this same rate, what would be the tax on a piece of property valued at $22,000?

Answer _____

5. A train is traveling at the rate of 75 miles per hour. At this speed, how far will it travel in 40 minutes? (Change 1 hour to minutes.)

Answer _____

6. Terry bought 3 yards of fabric for $12.75. How much would 5 yards of the same fabric have cost?

Answer _____

Variation

In the formula $D = rt$, distance equals the rate times the time. If a car travels at a rate of r miles per hour for 3 hours, the distance D that it travels can be found by using the equation $D = r \times 3$. If the car doubles its time, it will double its distance. If the car triples its time it will triple its distance. The distance traveled **varies directly** as the time. This is an example of **direct** variation.

The area A of a circle with radius r is given by $A = \pi r^2$. If the radius doubles to $2r$, the area becomes $\pi (2r)^2 = \pi \times 2r \times 2r = \pi (4r^2)$. Thus, the area of a circle varies directly as the *square* of the radius.

PRACTICE

Complete the tables.

1. $D = rt$

r	50	50	50	50	50
t	1	2	3	4	5
D	50				

Distance varies directly as the time.

2. $A = lw$

w	10	10	10	10	10	10	10
l	2	4	6	8	10	12	14
A	20						

Area varies directly as the length.

3. $A = \pi r^2$

r	7	14	21	28	35	42	49
A	154						

Area varies directly as the square of the radius. Use $\pi = \frac{22}{7}$.

PROBLEM-SOLVING STRATEGY

Use a Formula

The facts in a problem may be related by a formula. The chart lists some commonly used formulas.

$$\text{Area of a triangle:} \quad A = \frac{1}{2}bh$$
$$\text{Volume of a rectangular prism:} \quad V = lwh$$
$$\text{Simple interest:} \quad I = prt$$

EXAMPLE 1

Read the problem.

A triangle has a base *(b)* of 8 inches and a height *(h)* of 6 inches. What is the area of the triangle?

Select a formula.

Use the formula for the area of a triangle.
$A = \frac{1}{2}bh$

Substitute the data.

Replace the *b* and *h* with the numbers given.
$A = \frac{1}{2}b \times h$
$A = \frac{1}{2} \times 8 \times 6$

Solve the problem.

$A = \frac{1}{2} \times 8 \times 6$
$A = 24$
The area of the triangle is 24 *square* inches.

EXAMPLE 2

Read the problem.

The volume *(V)* of a rectangular prism is 288 cubic inches. The length *(l)* is 9 inches and the width *(w)* is 4 inches. What is the height *(h)*?

Select a formula.

Use the formula for the volume of a prism.
$V = lwh$

Substitute the data.

$V = l \times w \times h$
$288 = 9 \times 4 \times h$

Solve the problem.

$288 = 9 \times 4 \times h$
$288 = 36 \times h$
$288 \div 36 = h$
$8 = h$
The height of the prism is 8 inches.

Solve. Use one of the formulas given in the chart.

1. Chan invested $2000 (p) in a savings account paying an interest rate (r) of 3% annually. Find the interest (I) he will earn in a time (t) of 8 years.

   Answer _____

2. Louisa had a savings account that paid 2.5% interest annually. Her $1500 investment earned $225 in interest. For how many years was the money invested?

   Answer _____

3. The volume *(V)* of a rectangular prism is 84 cubic inches. The height *(h)* of the prism is 3 inches. The length *(l)* is 7 inches. What is the width of the prism?

   Answer _____

4. The area *(A)* of a triangle is 12 square inches. The height *(h)* of the triangle is 4 inches. What is the base *(b)* of the triangle?

   Answer _____

5. One triangular face of a pyramid has a base *(b)* of 20 meters and a height *(h)* of 30 meters. What is the area of the triangular face?

   Answer _____

6. A triangular stained glass window has a base of 36 inches and a height of 48 inches. What is the area of the window?

   Answer _____

7. A wastebasket has a length *(l)* of 20 inches, a width of 10 inches, and a height *(h)* of 24 inches. What is the volume of the wastebasket?

   Answer _____

8. A mailing box has a length of 18 inches, a width of 20 inches, and a height of 5 inches. What is the volume of the box?

   Answer _____

SOLVING AND USING EQUATIONS

Similar Triangles

Two triangles are *similar* if the corresponding angles are the same size and if the ratio of the lengths of corresponding sides are equal. Since the ratios are equal, they form a proportion.

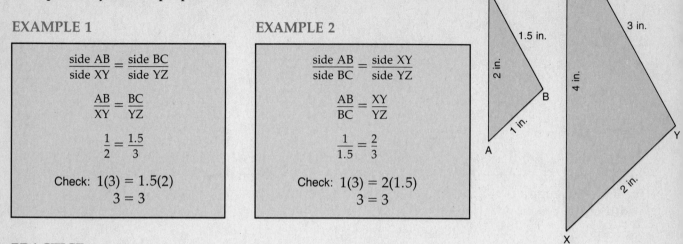

EXAMPLE 1

$$\frac{\text{side AB}}{\text{side XY}} = \frac{\text{side BC}}{\text{side YZ}}$$

$$\frac{AB}{XY} = \frac{BC}{YZ}$$

$$\frac{1}{2} = \frac{1.5}{3}$$

Check: $1(3) = 1.5(2)$

$3 = 3$

EXAMPLE 2

$$\frac{\text{side AB}}{\text{side BC}} = \frac{\text{side XY}}{\text{side YZ}}$$

$$\frac{AB}{BC} = \frac{XY}{YZ}$$

$$\frac{1}{1.5} = \frac{2}{3}$$

Check: $1(3) = 2(1.5)$

$3 = 3$

PRACTICE

Complete these proportions for the similar triangles above.

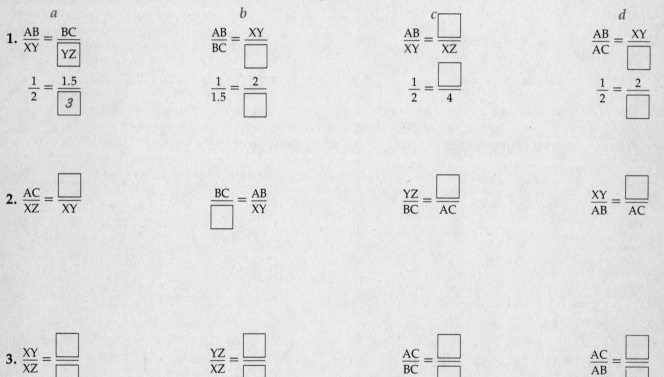

a

1. $\dfrac{AB}{XY} = \dfrac{BC}{\boxed{YZ}}$

$\dfrac{1}{2} = \dfrac{1.5}{\boxed{3}}$

b

$\dfrac{AB}{BC} = \dfrac{XY}{\boxed{}}$

$\dfrac{1}{1.5} = \dfrac{2}{\boxed{}}$

c

$\dfrac{AB}{XY} = \dfrac{\boxed{}}{XZ}$

$\dfrac{1}{2} = \dfrac{\boxed{}}{4}$

d

$\dfrac{AB}{AC} = \dfrac{XY}{\boxed{}}$

$\dfrac{1}{2} = \dfrac{2}{\boxed{}}$

2. $\dfrac{AC}{XZ} = \dfrac{\boxed{}}{XY}$

$\dfrac{BC}{\boxed{}} = \dfrac{AB}{XY}$

$\dfrac{YZ}{BC} = \dfrac{\boxed{}}{AC}$

$\dfrac{XY}{AB} = \dfrac{\boxed{}}{AC}$

3. $\dfrac{XY}{XZ} = \dfrac{\boxed{}}{\boxed{}}$

$\dfrac{YZ}{XZ} = \dfrac{\boxed{}}{\boxed{}}$

$\dfrac{AC}{BC} = \dfrac{\boxed{}}{\boxed{}}$

$\dfrac{AC}{AB} = \dfrac{\boxed{}}{\boxed{}}$

SOLVING AND USING EQUATIONS
Similar Triangles

If you are given the lengths of some of the sides of two similar triangles, you can find the lengths of other sides. To do this, set up and solve proportions.

In similar triangles ABC and XYZ, AB = 5 ft, XY = 6 ft, and BC = 10 ft. What is the length of YZ?

$$\frac{AB}{XY} = \frac{BC}{YZ} \qquad \text{or} \qquad \frac{AB}{BC} = \frac{XY}{YZ}$$

$$\frac{5}{6} = \frac{10}{YZ} \qquad\qquad \frac{5}{10} = \frac{6}{YZ}$$

$$5YZ = 60 \qquad\qquad 5YZ = 60$$

$$YZ = 12 \qquad\qquad YZ = 12$$

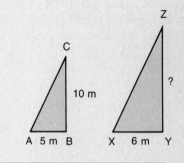

Set up and solve a proportion to solve each problem.

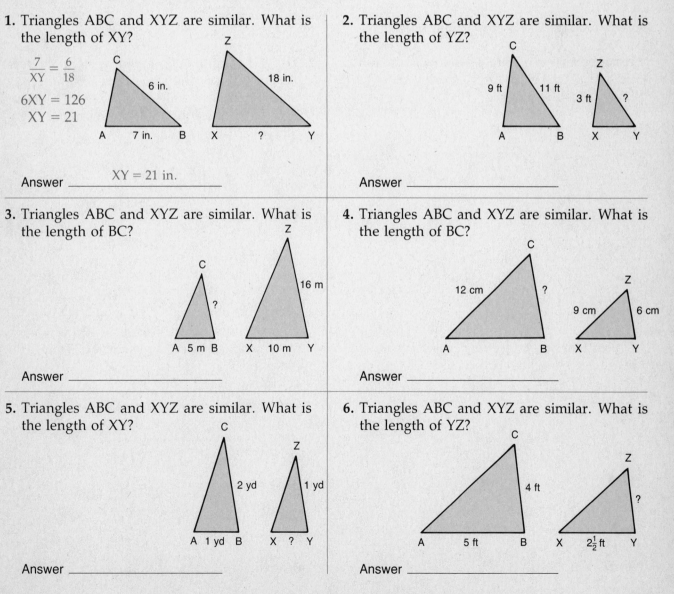

1. Triangles ABC and XYZ are similar. What is the length of XY?

$$\frac{7}{XY} = \frac{6}{18}$$

$$6XY = 126$$

$$XY = 21$$

Answer _____ XY = 21 in. _____

2. Triangles ABC and XYZ are similar. What is the length of YZ?

Answer _____

3. Triangles ABC and XYZ are similar. What is the length of BC?

Answer _____

4. Triangles ABC and XYZ are similar. What is the length of BC?

Answer _____

5. Triangles ABC and XYZ are similar. What is the length of XY?

Answer _____

6. Triangles ABC and XYZ are similar. What is the length of YZ?

Answer _____

197

SOLVING AND USING EQUATIONS
Using Proportion in Similar Figures

You can use proportions to solve problems about similar figures.

To find the height of the tree shown here, use similar triangles to form a proportion. Then solve.

A tree casts a shadow 60 feet long. At the same time, an 8-foot post casts a 12-foot shadow. How tall is the tree?

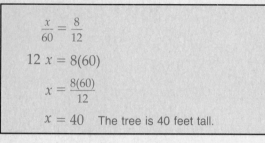

$$\frac{x}{60} = \frac{8}{12}$$

$$12\,x = 8(60)$$

$$x = \frac{8(60)}{12}$$

$$x = 40 \quad \text{The tree is 40 feet tall.}$$

PRACTICE

Draw a picture. Use a proportion. Solve.

1. A flagpole casts a 75-foot shadow at the same time a tree 20 feet tall casts a shadow of 30 feet. How tall is the flagpole?

Answer _____

2. The Washington Monument casts a shadow 111 feet long at the same time a 50-foot telephone pole casts a shadow 10 feet long. How high is the Washington Monument?

Answer _____

3. A telephone pole casts a shadow 30 feet long while a fence post 4 feet high casts a shadow 3 feet long. How high is the pole?

Answer _____

4. A smokestack casts a shadow of 40 feet while a fence post nearby casts a shadow of 2 feet. The fence post is 5 feet high. How tall is the smokestack?

Answer _____

PROBLEM SOLVING

Applications

Solve.

1. Ana has $2.50. Celia has $2.00. What is the ratio of Ana's money to Celia's? What is the ratio of Celia's to Ana's?

   Answer _____

   Answer _____

2. Ted, Bill, and Cheryl together have $84. Bill has twice as much as Ted. Cheryl has twice as much as Bill. How much does each have?

   Ted _____

   Bill _____

   Cheryl _____

3. A town has a population of 40,000 persons. There are 6000 school-age children in the town. What is the ratio of the school children to the total population?

   Answer _____

4. Daniel is 3 years older than twice Anita's age. The sum of their ages is 63 years. How old is each?

   Daniel _____

   Anita _____

5. Mr. Franklin is one third as old as his father. The sum of their ages is 100 years. How old is each person?

   Mr. Franklin _____

   Mr. Franklin's father _____

6. Gregory earned $52 in 12 hours. Use a proportion to show how much he would earn in 15 hours.

   Answer _____

7. Mr. Garcia and his daughter opened a business. Mr. Garcia invested 3 times as much money as his daughter. Together they invested $80,000. How much did each invest?

   Mr. Garcia _____

   Mr. Garcia's daughter _____

8. Tara has $15 more than twice as much as Josh does. Together, they have $45. How much does each have?

   Tara _____

   Josh _____

PROBLEM-SOLVING STRATEGY

Select a Strategy

In this book you have worked with several different problem-solving strategies. Some of them are listed in the box at the right.

PROBLEM-SOLVING STRATEGIES

Choose an Operation

Make a Drawing

Make a List

Use Logic

Use Estimation

Use a Formula

Use Guess and Check

Work Backwards

Read the problem. Select a strategy from the box. Then solve.

1. A marathon runner has been running for one and one-half hours at 11.6 miles per hour. The race is 26.2 miles long. How much farther does the runner have to run?

   Answer _____

2. Quality Cab charges $1.50 for the first mile and $0.30 for each additional $\frac{1}{4}$ mile. Comfy Cab charges $2.40 for the first mile and $0.20 for each additional $\frac{1}{4}$ mile. At what distance do the two charge the same?

   Answer _____

3. A spider was crawling to its nest on the side of a 650-foot skyscraper. One day the spider crawled up 120 feet, slipped back 22 feet, crawled up 90 feet, slipped back 16 feet, and then crawled the final 52 feet to the nest. How far above the ground is the spider's nest?

   Answer _____

4. The sum of the ages of Diane and her younger brother is 18. The product of their ages is 45. How old is each person?

   Diane _____

   Diane's brother _____

5. Braided rugs are offered in 4 basic colors, 3 different sizes, and 3 different styles. How many different color-size-style combinations are possible?

Answer _____

6. A rectangular picture has a width of 48 centimeters. The area of the picture is 3456 square centimeters. How long is the picture?

Answer _____

7. The number of people attending the Spartans basketball games has decreased steadily over the past few years. Two years ago the average attendance was 811 people. Last year the average attendance was 762. This year the average attendance was 713. If this trend continues, what will the average attendance be next year?

Answer _____

8. Lou deposited $1500 in a savings account. The interest rate was 3.75% annually. How much interest will he earn in 5 years?

Answer _____

9. Jan, Blake, and Gayle live in Richmond, Brookdale, and Montague but not necessarily in that order. Blake did not choose a home in Richmond. Neither Blake nor Jan live in Brookdale. Which person lives in which town?

Richmond _____

Brookdale _____

Montague _____

10. Tanya wants to install pine paneling on a wall in her den. The wall is 18 feet long and $8\frac{1}{2}$ feet high. Paneling costs $3.80 per square foot. How much will the paneling cost?

Answer _____

201

PROBLEM SOLVING

Applications

Solve.

1. Lou had a piece of cloth that was 6 meters long. He cut the cloth into pieces that measured 0.5 meters. How many pieces did he get from the original piece?

Answer _____

2. In an election, $\frac{3}{4}$ of the voters chose Marie De La Cruz. There were 8000 ballots cast. How many votes did she receive?

Answer _____

3. Bea got a 5% discount on a used car. The original price of the car was $5700. How much did her discount save her?

Answer _____

4. The area of a triangle is 80 square feet. The base of the triangle measures 20 feet. What is the height of the triangle?

Answer _____

5. How many yards of fence does Raul need to fence in a garden that is $12\frac{2}{3}$ yards long and 9 yards wide?

Answer _____

6. Jeanne has twice as much money as Jeff. Together they have $126. How much money does each person have?

Jeanne _____

Jeff _____

7. The sides of an enlarged photo are twice the length of the sides of the original photo. How much greater is the area of the enlarged photo than the area of the original?

Answer _____

8. A parking lot has an area of 85,000 square feet. The length of the parking lot is 340 feet. How wide is the parking lot?

Answer _____

PROBLEM SOLVING

Applications

Solve.

1. The area of one rectangle is twice the area of a second rectangle. The sum of their areas is 600 square meters. What is the area of each rectangle?

Rectangle 1 _____

Rectangle 2 _____

2. One of the fastest growing cities in the United States had a population of 587,720 in 1960. The 1970 census listed the population as 112.5% of the 1960 population. How large was the city in 1970?

Answer _____

3. The Empire State Building casts a shadow 250 feet long while a nearby 100-foot flagpole casts a shadow 20 feet long. How tall is the Empire State Building?

Answer _____

4. A business was allowed a $\frac{1}{2}$% discount for prompt payment of a bill for $800. What was the amount of discount?

Answer _____

5. The record for one day's rainfall for a city is 11.17 inches. The next heaviest is 8.32 inches. How much greater is the record rainfall?

Answer _____

6. A cord of wood is a stack 8 feet long, 4 feet wide, and 4 feet high. How many cubic feet are in a cord?

Answer _____

7. Thomas shoveled $12\frac{1}{2}$ cubic yards of sand in $2\frac{1}{2}$ hours. At that rate, how many cubic yards of sand did he shovel in one hour?

Answer _____

8. Last year, a salesman sold $160,500 worth of merchandise. His commission was 12%. How much was his commission?

Answer _____

SOLVING AND USING EQUATIONS

Unit 8 Review

Solve. Simplify.

	a	*b*	*c*	*d*
1.	$x + 15 = 23$	$3x = 63$	$8x + 5 = 29$	$6x - 9 = 21$

2. $4x + 2x = 48$ $\qquad$ $20x - 4x = 64$ $\qquad$ $7x - 2x = 80$ $\qquad$ $6x + 2 = 38 + 2x$

3. $3x + 6 = 41 - 2x$ $\qquad$ $7x = 24$ $\qquad$ $x + 1\frac{3}{8} = 5\frac{1}{4}$ $\qquad$ $x - \frac{2}{5} = \frac{3}{5}$

4. $\frac{9}{7} = \frac{x}{21}$ $\qquad$ $\frac{x}{50} = \frac{3}{25}$ $\qquad$ $\frac{5}{7} = \frac{x}{42}$ $\qquad$ $\frac{3}{x} = \frac{18}{30}$

Solve.

5. Lynn bought two extension cords. The second was twice as long as the first. In all, they were 48 feet long. How long was each cord?

Cord 1 _____

Cord 2 _____

6. A train can travel 210 miles in 5 hours. If it keeps the same speed, how far will it travel in 6 hours?

Answer _____

7. The sides of a cube are 10 inches long. The sides of a second cube are 3 times that long. How many times greater is the volume of the second cube than the volume of the first?

Answer _____

8. Triangle ABC is similar to triangle XYZ. What is the length of XY?

Answer _____

FINAL REVIEW

Write the value of the underlined digit.

	a	b	c	d
1.	78,309 _____	308 _____	7.215 _____	1.006 _____

Find each answer.

	a	b	c	d
2.	8098 +1456	23,067 + 8,692	659 −193	3870 −1779

3.	36 ×24	825 ×173	18)469	9)828

Line up the digits. Then find each answer.

	a	b	c
4.	294 − 41 = _____	785 + 928 = _____	11,382 − 807 = _____
5.	27 × 316 = _____	168 ÷ 3 = _____	2543 ÷ 62 = _____

Simplify.

	a	b	c	d	e
6.	$\frac{6}{15}=$	$\frac{12}{8}=$	$\frac{12}{36}=$	$\frac{42}{9}=$	$\frac{6}{50}=$

Solve.

7. The storekeeper sold $9\frac{1}{4}$ yards from a piece of material containing 53 yards. How much was left in the original piece?

8. Oranges sell for 46¢ per pound. How much will Ted have to pay for five and one half pounds?

Answer _____

Answer _____

Add or subtract. Simplify.

	a	*b*	*c*	*d*
9.	$\frac{3}{8}$	$\frac{5}{6}$	$\frac{3}{4}$	$\frac{5}{12}$
	$+\frac{3}{8}$	$+\frac{1}{6}$	$+\frac{1}{2}$	$+\frac{3}{8}$
10.	$\frac{7}{8}$	$\frac{1}{3}$	$1\frac{1}{2}$	$7\frac{8}{15}$
	$-\frac{1}{8}$	$-\frac{1}{10}$	$+2\frac{5}{6}$	$-1\frac{1}{5}$
11.	$7\frac{2}{3}$	$6\frac{3}{10}$	$9\frac{5}{8}$	2
	$+9\frac{1}{5}$	$-4\frac{4}{5}$	-7	$-1\frac{2}{3}$

Multiply or divide. Use cancellation when possible. Simplify.

	a	*b*	*c*
12.	$\frac{1}{2} \times \frac{2}{5} =$	$\frac{2}{9} \times \frac{3}{5} =$	$\frac{7}{8} \times \frac{4}{9} =$
13.	$\frac{5}{6} \times 12 =$	$\frac{3}{8} \times 10 =$	$\frac{1}{6} \times \frac{3}{7} =$
14.	$1\frac{3}{5} \times 10 =$	$1\frac{2}{3} \times 2\frac{1}{4} =$	$4\frac{2}{5} \times 1\frac{4}{11} =$
15.	$\frac{3}{5} \div \frac{2}{5} =$	$\frac{7}{8} \div 7 =$	$\frac{2}{3} \div \frac{1}{4} =$
16.	$2\frac{1}{4} \div 1\frac{1}{2} =$	$3\frac{3}{10} \div \frac{2}{5} =$	$\frac{3}{5} \div 1\frac{1}{5} =$
17.	$2\frac{3}{4} \div \frac{11}{15} =$	$3\frac{1}{4} \div 4\frac{1}{3} =$	$\frac{1}{5} \div \frac{5}{9} =$

Write each decimal as a fraction.

	a	*b*	*c*	*d*
18.	0.3 = _____	1.17 = _____	0.009 = _____	0.01 = _____

Write each fraction as a decimal.

	a	*b*	*c*	*d*
19.	$\frac{8}{10}$ = _____	$\frac{23}{100}$ = _____	$\frac{5}{100}$ = _____	$\frac{8}{1000}$ = _____

Find each answer. Write zeros as needed.

	a	*b*	*c*	*d*
20.	7.3 8 +0.1 9	4 8.0 2 +2 1.0 8 6	8.9 5 −1.6 8	1 5.3 − 7.2 9

| **21.** | 9.0 6
× 1 3 | 0.0 0 1 4
× 2 4 9 | 8.3 4
× 5 | 7.6 9
×0.1 3 |

22.

| $10\overline{)7.9}$ | $0.1\overline{)1.6\,5}$ | $9\overline{)2\,2.7\,7}$ | $1.5\overline{)1\,0.0\,5}$ |

Write each percent as a decimal and as a fraction.

	a	*b*
23.	35% = _____ _____	8% = _____ _____

Find each number.

	a	*b*
24.	25% of 80	What percent of 40 is 8?
25.	10% of what number is 9?	72% of 50

Solve.

26. A triangle has a base of 9 meters and a height of 12 meters. What is the area?

27. How much linoleum is needed to cover a rectangular floor that measures 10 feet by $12\frac{1}{2}$ feet?

Answer _____

Answer _____

Change each measurement to the unit given.

	a	b	c
28.	10 qt = _____ gal _____ qt	8 pt = _____ c	12 oz = _____ lb
29.	16 in. = _____ ft _____ in.	1.5 T = _____ lb	6 mL = _____ L
30.	4 kg = _____ g	230 mg = _____ g	9000 g = _____ kg

Solve.

	a	b	c	d
31.	$x - 15 = 7$	$x + 5 = 19$	$7x = 42$	$6x + 1 = 61$
32.	$3x + 5x = 72$	$10x - 4x = 36$	$\dfrac{5}{x} = \dfrac{40}{64}$	$\dfrac{12}{5} = \dfrac{x}{15}$
33.	$\dfrac{x}{9} = \dfrac{12}{15}$	$6x + 9 = 29 + 2x$	$5x - 1 = 24$	$6 = 2x - 20$
34.	$\dfrac{7}{5} = \dfrac{x}{10}$	$x - 50 = 30$	$17 = 2 + x$	$10x = 5x + 15$

Write the fraction for each ratio.

	a	b
35.	The ratio of days in a week to months in a year	The ratio of 3 teachers to 24 students
	Ratio: _____	Ratio: _____

Solve the proportion.

	a	b	c
36.	$\dfrac{3}{8} = \dfrac{x}{40}$	$\dfrac{2}{x} = \dfrac{9}{18}$	$\dfrac{x}{6} = \dfrac{5}{12}$